SpringerBriefs in Mathematics

SpringerBriefs in Mathematics showcases expositions in all areas of mathematics and applied mathematics. Manuscripts presenting new results or a single new result in a classical field, new field, or an emerging topic, applications, or bridges between new results and already published works, are encouraged. The series is intended for mathematicians and applied mathematicians.

Titles from this series are indexed by Web of Science, Mathematical Reviews, and zbMATH.

More information about this series at http://www.springer.com/series/10030

Daniel Alpay · Fabrizio Colombo ·
Irene Sabadini

Quaternionic de Branges Spaces and Characteristic Operator Function

Springer

Daniel Alpay
Schmid College of Science and Technology
Chapman University
Orange, CA, USA

Fabrizio Colombo
Dipartimento di Matematica
Politecnico di Milano
Milano, Italy

Irene Sabadini
Dipartimento di Matematica
Politecnico di Milano
Milano, Italy

ISSN 2191-8198   ISSN 2191-8201   (electronic)
SpringerBriefs in Mathematics
ISBN 978-3-030-38311-4   ISBN 978-3-030-38312-1   (eBook)
https://doi.org/10.1007/978-3-030-38312-1

Mathematics Subject Classification (2010): 46E22, 47S10, 30G35, 46C20

This Springer imprint is published by the registered company Springer Nature Switzerland AG
The registered company address is: Gewerbestrasse 11, 6330 Cham, Switzerland

Preface

This work inserts in the very fruitful study of quaternionic linear operators. This study is a generalization of the complex case, but the noncommutative setting of quaternions shows several interesting new features, see, e.g., the so-called S-spectrum and S-resolvent operators. In this work, we study de Branges spaces, namely, the quaternionic counterparts of spaces of analytic functions (in a suitable sense) with some specific reproducing kernels, in the unit ball of quaternions or in the half-space of quaternions with positive real parts. The spaces under consideration will be Hilbert or Pontryagin or Krein spaces. These spaces are closely related to operator models that are also discussed. The focus of this book is the notion of characteristic operator function of a bounded linear operator A with finite real part, and we address several questions like the study of J-contractive functions, where J is self-adjoint and unitary, and we also treat the inverse problem, namely, to characterize which J-contractive functions are characteristic operator functions of an operator. In particular, we prove the counterpart of Potapov's factorization theorem in this framework. Besides other topics, we consider canonical differential equations in the setting of slice hyperholomorphic functions and we define the lossless inverse scattering problem. We also consider the inverse scattering problem associated with canonical differential equations. These equations provide a convenient unifying framework to discuss a number of questions pertaining, for example, to inverse scattering, non-linear partial differential equations and are studied in the last section of this book.

A problem which is related to our study, which is one of the crucial problems in operator theory, is determining the invariant subspaces of a linear closed operator. Working on Hilbert spaces, the spectral theorem for normal operators is one of the most important achievements of the last century. Even though there has been a lot for works regarding the problem of extending the reduction theory to non-normal linear operators, still a lot of problems are unsolved nowadays.

In order to study the spectral analysis for a number of non-self-adjoint operators one has to extend the theory to non-normal operators in a Hilbert space and to operators in Banach spaces; this has been done with the theory of spectral operators, see [65].

An important contribution to the reduction theory of non-normal operators was the notion of characteristic operator function introduced by Livsic.

In the quaternionic setting things are much more complicated since the appropriate notion, namely, the S-spectrum of a quaternionic linear operator was introduced 70 years after the paper of Birkhoff and von Neumann on the logic of quantum mechanics that was published in 1936. Moreover, the spectral theorem for quaternionic normal linear operators (bounded or unbounded) was proved in 2015 and appeared in the literature in 2016. We recall that, if T is a bounded linear quaternionic operator then the S-spectrum is defined as

$$\sigma_S(T) = \{s \in \mathbb{H} \ : \ T^2 - 2\mathrm{Re}(s)T + |s|^2 I \text{ is not invertible}\},$$

while the S-resolvent set is

$$\rho_S(T) := \mathbb{H} \backslash \sigma_S(T).$$

Let us restrict to that case when T is a bounded normal quaternionic linear operator on a quaternionic Hilbert space $\mathcal{H}$. Then there exist three quaternionic linear operators A, J, B such that $T = A + JB$, where A is self-adjoint and B is positive, J is an anti-self-adjoint partial isometry (called imaginary operator). Moreover, A, B and J mutually commute.

Let us set $\mathbb{C}_j^+ = \{u + jv, (u, v) \in \mathbb{R} \times \mathbb{R}^+\}$, for $j \in \mathbb{S}$, where $\mathbb{S}$ is the unit sphere of purely imaginary quaternions. So the spectral theorem is as follows. There exists a unique spectral measure E_j on $\sigma_S(T) \cap \mathbb{C}_j^+$ so that for any slice continuous intrinsic function $f = f_0 + f_1 j$ and $x, y \in \mathcal{H}$

$$\langle f(T)x, y \rangle = \int_{\sigma_S(T) \cap \mathbb{C}_j^+} f_0(q) \, d\langle E_j(q)x, y \rangle + \int_{\sigma_S(T) \cap \mathbb{C}_j^+} f_1(q) \, d\langle JE_j(q)x, y \rangle. \quad (1)$$

With the spectral theorem and the S-functional calculus, that is the quaternionic analogue of the Riesz-Dunford functional calculus, it turned out to be clear that to replace complex spectral theory with quaternionic spectral theory we have to replace the classical spectrum with the S-spectrum.

The first direction of research of operator theory in the quaternionic setting, beyond the spectral theorem based on the S-spectrum, was done in the recent long paper [73], where the author studies quaternionic spectral operators. The present book considers a different avenue and begins the investigation of the quaternionic characteristic operator function. In classical operator theory the notion of resolvent operator plays a key role. More precisely, let T be a possibly unbounded linear operator acting on a Hilbert space $\mathcal{H}$. The resolvent operator $R(z) = (T - zI_{\mathcal{H}})^{-1}$ is an operator-valued function, analytic on the resolvent set $\rho(T)$ of T, assumed non-empty, and its properties and those of T are closely related. Characteristic operator functions are possibly simpler analytic functions, built on $R(z)$, and still allowing to deduce properties of the operator from properties of the functions. The characteristic operator function associated with a close-to-unitary operator originates with the work of Livsic; see [93]. Properties and applications of the characteristic operator function of an operator which is close-to-self-adjoint are discussed in particular in the book [51].

This work consists of ten chapters, the Preliminaries being the first. Chapters 2–4 may be seen of a preliminary nature, although they also contain some new material. In Chapter 2 we recall some facts on quaternions, quaternionic matrices and quaternionic functional analysis. The main aspects needed in this work on the theory of slice hyperholomorphic functions as well as on the S-resolvent operators and the S-spectrum are recalled in Chapter 3. The original part in the section consists in the study of slice hyperholomorphic weights, both in the case of the unit ball and of the half-space. A key tool is a map, denoted by ω, which allows to rewrite the values of a quaternionic valued function in terms of 2×2 matrices with complex entries.

The theory of slice hyperholomorphic rational functions and their symmetries is considered in Chapter 4. Operator models in the sense of Rota are studied in Chapter 5. In Chapter 6 we consider quaternionic $\mathcal{H}(A, B)$ spaces and we provide the counterparts of various results in this framework, including the operator of multiplication in the half-space case and in the unit ball case and the study of the reproducing kernels. The case of J-contractive functions is presented in Chapter 7. The characteristic operator function is defined and studied in Chapter 8, where we also provide examples and we discuss inverse problems. Some classes of functions with a positive real part in the half-space or the unit ball are studied in Chapter 9. Finally, Chapter 10 is devoted to the canonical differential systems in the quaternionic setting, also those associated with an operator and, in particular, we study the matrizant.

We consider this manuscript as a seminal work which can be expanded and can give rise to several different directions of research in operator theory and hypercomplex analysis.

Orange, USA   Daniel Alpay
Milano, Italy   Fabrizio Colombo
Milano, Italy   Irene Sabadini

Contents

Chapter 1
Preliminaries

1.1 The Complex Numbers Setting

Since the purpose of this work is to study the characteristic operator function and related topics in the quaternionic setting and for the sake of completeness, we recall in this chapter some definitions from the classical complex case.

Let T be a bounded linear operator in a Hilbert space $\mathcal{H}$ over the complex numbers, with finite dimensional imaginary part, meaning that the operator $\frac{T-T^*}{2i}$ has a finite dimensional everywhere defined extension (say of rank n), which we write as

$$\frac{T-T^*}{2i} = KJK^*, \tag{1.1}$$

where $J \in \mathbb{C}^{n \times n}$, ($\mathbb{C}^{n \times n}$ denotes the set of $n \times n$ matrices with complex entries), is both self-adjoint and unitary (i.e., J is a signature matrix) and where K is a linear bounded operator from $\mathbb{C}^n$ into $\mathcal{H}$. The characteristic operator function of the operator T is then defined by

$$W(z) = I - 2iK^*(T - zI)^{-1}KJ, \tag{1.2}$$

see [51]. Note that the imaginary part of T may be infinite dimensional in [51], while the present work focuses in the finite dimensional case, namely, in the quaternionic context, on operators with finite dimensional real part.

The function W is analytic in $\mathbb{C} \setminus \sigma(T)$, where we denoted by $\sigma(T)$ the spectrum of T; it is J-expansive in the open upper half-plane $\mathbb{C}_+$ and J-contractive in the open lower half-plane $\mathbb{C}_-$, namely,

$$\begin{aligned} W(z)^*JW(z) &\geq J, \quad z \in \rho(T) \cap \mathbb{C}_+, \\ W(z)^*JW(z) &\leq J, \quad z \in \rho(T) \cap \mathbb{C}_-. \end{aligned} \tag{1.3}$$

Note that often, for a given J, one considers functions J-contractive rather than J-expansive in $\mathbb{C}_+$.

© The Author(s), under exclusive license to Springer Nature Switzerland AG 2020
D. Alpay et al., *Quaternionic de Branges Spaces and Characteristic
Operator Function*, SpringerBriefs in Mathematics,
https://doi.org/10.1007/978-3-030-38312-1_1

The study of the relationships between the properties of the function $W(z)$ and of the operator T leads to a number of important problems, of which we mention in particular:

- Relate the spectrum of T and the singularities of W.
- Relate factorizations of W and invariant subspaces of T.
- Inverse problem: when is a J-contractive function the characteristic operator of some operator?
- The indefinite setting case, where the Hilbert space is replaced by a Pontryagin space, or possibly by a Krein space.

Operator models are closely related to Hilbert (and Pontryagin) spaces of analytic functions of different kinds, and were introduced in a series of work by de Branges and de Branges together with Rovnyak, see, e.g., [46, 48–50]. In particular, spaces with reproducing kernel of one of the following forms (and their counterparts with denominator equal to $1 - z\overline{w}$) play an important role:

(a) $\mathcal{H}(A, B)$ spaces, with reproducing kernel

$$\frac{A(z)A(w)^* - B(z)B(w)^*}{z + \overline{w}}, \tag{1.4}$$

where A and B are $\mathbb{C}^{n \times n}$-valued, the set of $n \times n$ matrices with entries in $\mathbb{C}$ and analytic in some open subset of the open right half-plane $\mathbb{C}_r$, with $\det A(z) \not\equiv 0$. When $S = A^{-1}B$ extends to an inner function (i.e., the boundary values are almost everywhere unitary), one has that $\mathcal{H}(A, B) = A\,(\mathbf{H}_2(\mathbb{C}_r) \ominus S\mathbf{H}_2(\mathbb{C}_r))$, where $\mathbf{H}_2(\mathbb{C}_r)$ is the Hardy space on $\mathbb{C}_r$, and the operator of multiplication by the variable is an Hermitian operator. The description of its self-adjoint extensions is a problem of interest and with, for instance, applications to interpolation problems (see [11] for the latter). We point out that it is sometimes easier to rewrite (1.4) as

$$\frac{E_+(z)E_+(w)^* - E_-(z)E_-(w)^*}{z + \overline{w}}, \tag{1.5}$$

with

$$E_+(z) = \frac{A(z) + B(z)}{\sqrt{2}} \quad \text{and} \quad E_-(z) = \frac{A(z) - B(z)}{\sqrt{2}}.$$

(b) $\mathcal{H}(\Theta)$ spaces, with reproducing kernel

$$\frac{J - \Theta(z)J\Theta(w)^*}{z + \overline{w}}, \tag{1.6}$$

where Θ is $\mathbb{C}^{2n \times 2n}$-valued and analytic in some open subset of the open right half-plane. Let

$$J_1 = \begin{pmatrix} 0 & I_n \\ I_n & 0 \end{pmatrix}, \tag{1.7}$$

and set $J = J_1$. Multiplying the left side of (1.6) by $(I_n \ 0)$ and by the transpose of this matrix on the left, we get the kernel

$$\frac{-\Theta_{11}(z)\Theta_{12}(w)^* + \Theta_{12}(z)\Theta_{11}(w)^*}{z + \overline{w}}$$

which is of the form (1.4), with

$$A(z) = \frac{\Theta_{11}(z) - \Theta_{12}(z)}{\sqrt{2}} \quad \text{and} \quad B(z) = \frac{\Theta_{11}(z) + \Theta_{12}(z)}{\sqrt{2}}.$$

The same argument can be made by multiplying the left side of (1.6) by $(0 \ I_n)$ and by the transpose of this matrix on the left. Also here we get a kernel of the form (1.4). In a number of cases there is a natural isometry between $\mathcal{H}(A, B)$ and $\mathcal{H}(\Theta)$, which is a generalization of the map sending orthogonal polynomials of the first kind to orthogonal polynomials of the second kind.

(c) $\mathcal{L}(\Phi)$ spaces, with reproducing kernel

$$\frac{\Phi(z) + \Phi(w)^*}{z + \overline{w}}, \tag{1.8}$$

where Φ is $\mathbb{C}^{n \times n}$-valued and analytic in some open subset of the open right half-plane (the associated reproducing kernel Hilbert spaces are models for self-adjoint operators and pairs of self-adjoint operators; see [36, 49]).

$\mathcal{L}(\Phi)$ spaces and $\mathcal{H}(\Theta)$ spaces are related by linear fractional transformations. More precisely, assume that in (1.6) we have $J = J_1$. A theorem of de Branges and Rovnyak, see [49], states that the map

$$F \mapsto (\Phi \ I_n) F$$

is a contraction from $\mathcal{H}(\Theta)$ into $\mathcal{L}(\Phi)$ if and only if one can write

$$\Phi = (\Theta_{22}\varphi - \Theta_{12})(\Theta_{11} - \Theta_{12}\varphi)^{-1},$$

where φ is analytic and has a real positive part in the left open half-plane. The directions in which φ is identically equal to ∞ have to be suitably interpreted. To avoid this problem one can rewrite the above linear fractional transformation as

$$\Phi = (\Theta_{22}(I_n + s) - \Theta_{21}(I_n - s))(\Theta_{11}(I_n - s) - \Theta_{12}(I_n + s))^{-1} \tag{1.9}$$

where s is analytic and contractive in $\mathbb{C}_r$, see [49, p. 306].

Definition 1.1.1. *The problem of finding all the linear fractional expressions* (1.9) *associated with a given function Φ is called the lossless inverse scattering problem.*

The term *lossless* in the above definition refers to the fact that Θ is usually assumed J-inner and is the chain-scattering matrix function of a lossless system. See [4] for more information.

The lossless inverse scattering problem allows to put under a common setting a wide range of questions. It was studied in particular in [25, 26]. When Θ is entire, it is the inverse spectral problem studied, in particular, in the books [41, 70].

Remarks 1.1.2. (1) Even though $\mathcal{H}(A, B)$ spaces can be formally related with $\mathcal{H}(\Theta)$ space by multiplicating their elements by the matrix $[I_n \ \ 0]$, it is not true that any $\mathcal{H}(A, B)$ space is the upper part of a $\mathcal{H}(\Theta)$ space. See [25, Theorem 3.1 p. 600] for such an embedding result for subspaces of $\mathbf{L}_2(d\mu)$, where $d\mu$ is a positive measure on the unit circle or on the real line, and see Theorem 3.4.3 in Chapter 3 for its quaternionic counterpart.

(2) We also note that finite dimensional $\mathcal{H}(A, B)$ spaces of polynomials are related to the Gohberg-Heinig and Christoffel-Darboux formulas; see Section 2.2.

(3) Finally, we remark that all the kernels defined above are of the form

$$\frac{X(z)\Sigma X(w)^*}{\rho_w(z)} \tag{1.10}$$

where Σ is a signature matrix, X is a matrix-valued analytic function of appropriate size and $\rho_w(z)$ is equal to either $z + \overline{w}$ or $1 - z\overline{w}$. The general theory of such spaces (for more general denominators) was initiated in [25] and we refer to [30, 66, 69] for further information.

A related important notion is that of canonical differential expressions. These are ordinary differential equations of the form

$$\frac{dG}{dt}(t, z) = izJH(t)G(t, z), \tag{1.11}$$

where z is a complex parameter, H is a given $\mathbb{C}^{2n \times 2n}$-valued function, $J \in \mathbb{C}^{2n \times 2n}$ is a signature matrix and the unknown G is a $\mathbb{C}^{2n \times m}$-valued function for some $m \in \mathbb{N}$. A simpler family of canonical differential expressions are differential equations of the form

$$iJ\frac{dF}{dt}(t, z) = (zI_{2n} + V(t))F(t, z), \tag{1.12}$$

where

$$J = \begin{pmatrix} I_n & 0 \\ 0 & -I_n \end{pmatrix} \quad \text{and} \quad V(t) = \begin{pmatrix} 0 & v(t) \\ v(t)^* & 0 \end{pmatrix},$$

and where the $\mathbb{C}^{n \times n}$-valued function v is called the potential. To see the connection between (1.11) and (1.12) consider the solution T of the equation

$$iJ\frac{dT}{dt}(t) = V(t)T(t)$$

where V is as in (1.12). Let F be a solution of (1.12) and define G by $F = TG$. Then,

$$iJ(T'G + TG') = (zI_{2n} + V)TG$$

and so G satisfies (1.11) with H defined by

$$H = -JT^{-1}JT.$$

It is not true that, conversely, any equation (1.11) leads to an equation (1.12) in such a way.

Associated with these canonical expressions there are a number of functions of z (among which we mention the scattering function, the Weyl function and the spectral function). Direct problems consist in computing these functions when H (or V in case (1.12)) are given, while inverse problems consist of recovering H (or V) from one of these functions. The study of these expressions forms a convenient framework to study a wide range of problems, including non-linear partial differential equations. See [102].

Remark 1.1.3. The connection between the theory of canonical differential equations and the notion of characteristic function, see (1.2), follows in particular from the fact that the solution to (1.12) subject to the initial condition $F(0, z) = I$ is J-expansive.

To provide a more complete description, it is useful to recall that the multiplicative structure of matrix-valued functions J-contractive in the open unit disc was given by Potapov in [95]. More precisely, he proved that any function Θ which is J-contractive in the unit disc can be written in a unique way (up to multiplicative constant factors) as a product

$$\Theta(z) = \Theta_1(z)\Theta_2(z)\Theta_3(z) \tag{1.13}$$

where Θ_1 is a (possibly infinite) Blaschke product analytic in $\mathbb{D}$, Θ_2 is a (possibly infinite) Blaschke product analytic in $|z| > 1$ and Θ_3 is J-contractive and with no zeros nor poles in $\mathbb{D}$. Both Θ_1 and Θ_2 are J-unitary on the unit circle. More precisely, let

$$b_a(z) = \frac{z - a}{1 - z\overline{a}},$$

where $a \in \mathbb{C}$ of modulus different from 1. The function Θ_1 (resp. Θ_2) is a (possibly infinite) product of terms of the form

$$\theta_{a,u}(z) = I + \left(\frac{b_a(z)}{b_a(1)} - 1\right)\frac{uu^*J}{u^*Ju} \tag{1.14}$$

where $|a| < 1$ and $u \in \mathbb{C}^n$ is such that $u^*Ju > 0$ (resp. $|a| > 1$ and $u \in \mathbb{C}^n$ is such that $u^*Ju < 0$), and called (normalized) Blaschke-Potapov factors of the first and second kind, respectively. Factors of the second kind will appear if and only if J has negative eigenvalues. In case of a rational J-unitary Θ, the function Θ_3 is a finite product of Blaschke-Potapov factors of the third kind, or Brune sections. When normalized to be identity at the point $z = 1$ (and in particular not to have a pole there), these are functions of the form

$$\left(I + e\frac{z+a}{z-a}uu^*J\right)\left(I + e\frac{1+a}{1-a}uu^*J\right)^{-1} = I + \frac{2e}{1-a}\frac{1-z}{z-a}uu^*J, \quad (1.15)$$

where now $|a| = 1, e > 0$ and $u^*Ju = 0$. We will refer to Θ_3 as to the singular factor. It is interesting to note that the function $\Theta_3(z)$ is expressed in terms of a multiplicative integral in the form

$$\Theta_3(z) = \overset{\frown}{\int_0^\ell} \exp\left(\frac{z + e^{i\theta(t)}}{z - e^{i\theta(t)}}dE(t)\right) \quad (1.16)$$

where $\theta(t)$ is increasing and $E(t)J$ is Hermitian, increasing (meaning that $E(t_1)J \leq E(t_2)J$, when $t_1 < t_2$) and normalized by imposing $\text{Tr}\,E(t)J = 1$.

We will refer to (1.13) as to the Potapov decomposition of the given function Θ. It is important to mention that the introduction of J contractive functions, via characteristic operator functions, was the main motivation for Potapov's work.

For completeness, we recall that the multiplicative integral $\overset{\frown}{\int_0^\ell} e^{f(t)dt}$ is defined as the limit of the products

$$\overset{\frown}{\prod_{j=0}^{k-1}} e^{f(s_j)(t_{j+1}-t_j)} = \overset{\frown}{\prod_{j=0}^{k-1}} (I + f(s_j)(t_{j+1} - t_j))$$

where $t_0 = 0 < t_1 < \cdots < t_k = \ell$ is a partition of $[0, \ell]$ and $s_j \in [t_j, t_{j+1}]$. The limit exists when f is continuous. We refer to the Appendix in Potapov's paper [95] for the basics on multiplicative integrals. In particular, a multiplicative integral

$$M(s) = \overset{\frown}{\int_0^s} e^{f(t)dt}$$

satisfies the following differential equation (see [95, Theorem p. 241]):

$$\frac{d}{ds}M(s) = M(s)f(s).$$

Finally, we recall that Potapov's decomposition leads to:

Theorem 1.1.4. *An entire function Θ which is J-inner in the open upper half-plane can be written as a multiplicative integral of the form*

$$\Theta(z) = \int_0^{\widetilde{\ell}} e^{-izH(t)dt} \Theta(0) \tag{1.17}$$

where $H(t)$ is integrable and such that $H(t)J \geq 0$ on $[0, \ell]$.

Furthermore, de Branges proved that H is unique when $n = 2$. We refer to [51] and to the book of Arov and Dym [42] for uniqueness conditions when $n > 2$.

We conclude this section with the following consequence of Theorem 1.1.4 which is of importance in several applications:

Corollary 1.1.5. *An entire function Θ J-inner in the open upper half-plane is of finite exponential type.*

1.2 The Quaternionic Setting

In this section we turn to the quaternionic case and we start by defining the quaternionic counterpart of characteristic operator function.

We note that there several substantial differences between the complex and the quaternionic case. First of all, linear spaces and linear operators can be considered on the left or on the right. These two cases are somewhat equivalent (in the sense that the statements which hold in one case hold also in the other) but different since the classes of right and of left linear operators are different. Moreover, we replace the imaginary line by the real line and the (complex) open upper half-plane by the (quaternionic) right half-space. Furthermore analytic functions and rational functions are replaced by slice hyperholomorphic functions and rational slice hyperholomorphic functions, and the pointwise product of analytic functions is replaced by the so-called $\star$-product defined in the next chapter.

Definition 1.2.1. *Let A be a continuous right linear operator in a right quaternionic space, with finite dimensional real part (say of rank n), and assume that*

$$A + A^* = -C^* JC, \tag{1.18}$$

where $J \in \mathbb{H}^{n \times n}$ (where $\mathbb{H}^{n \times n}$ is the set of $n \times n$ matrices with entries in $\mathbb{H}$) is both self-adjoint and unitary, and C is linear bounded from the quaternionic Hilbert space $\mathcal{H}$ into $\mathbb{H}^n$. The function

$$S(p) = I_n - pC^* \star (I - pA^*)^{-\star} CJ \tag{1.19}$$

is called the characteristic operator function of A.

We gave an analogous definition in [17] where, however, we considered A instead of its adjoint A^* in (1.19), but in view of the connections with canonical differential systems (see Section 10.1) it is more convenient to use the present definition. Note also the symmetry between A and A^* in (1.18). In the classical case, the operator and its adjoint have an anti-symmetric role rather than a symmetric role: replacing T by T^* in (1.1) changes J to $-J$.

In this work we wish to address the following questions in the setting of linear operators acting on quaternionic Hilbert spaces:

- What are the J-contractive functions.
- What is the analogue, if any, of (1.17).
- Which J-contractive functions are characteristic operator functions (inverse problem).
- How to associate to a given operator a canonical differential expression.
- What is the operator associated with a canonical differential expression (inverse problem).
- What is the lossless inverse scattering problem (see Definition 1.1.1).

Note that usually we will take J with real entries; in particular we set

$$J_0 = \begin{pmatrix} I_n & 0 \\ 0 & -I_n \end{pmatrix}, \qquad (1.20)$$

where I_n (often denoted by I) denotes the identity matrix of order n.

Remark 1.2.2. The approach and the methods in the present work and in our book [23] are completely different. There, the emphasis was on the notion of realization, and a key role was played by a result of Shmul'yan [109] on extensions of linear relations in Pontryagin spaces. In the present work the emphasis is on the spaces themselves and on their connections to underlying problems such as the lossless inverse scattering problem (see Definition 1.1.1 for the latter).

Remark 1.2.3. A recurring theme in the book, and in some of our previous works, is the following: we are given a $\mathbb{H}^{n \times n}$-valued function $K(p, q)$, positive definite in an axially symmetric open subset of the open unit ball of the quaternion, and slice hyperholomorphic on the left in p and on the right in $\bar{q}$ there (here $\mathbb{H}^{n \times n}$ denotes the set of $n \times n$ matrices with quaternionic entries). We restrict to $\Omega \cap \mathbb{R}$ and consider $p = x$ and $q = y$ real, then we apply the map χ (see (2.2)) to reduce to the case of matrix-valued functions whose entries are complex. For the quaternionic kernel $K(p, q)$, the obtained kernel is of the form

$$A(x) \frac{V(x) + V(y)^*}{1 - xy} A(y)^*, \qquad (1.21)$$

where A is, say $\mathbb{C}^{2n \times 2n}$-valued, analytic and invertible, and V is *defined* on $\Omega \cap \mathbb{R}$. Assume that the kernel is positive definite in $\Omega \cap \mathbb{R}$. Loewner's theorem (see [64, Theorem 1, p. 95]) or arguments using function theory in the Hardy space (see [4]) will allow us to assert that V extends to a function analytic in the open unit disc, and with a positive real part there. This allows us to use methods of complex analysis to study the original kernel $K(p, q)$, and the case where (1.21) is replaced by

$$A(x)\frac{V(x) + V(y)^*}{x + y} A(y)^*.$$

A weaker fact holds when the above kernels are assumed to have a finite number of negative squares rather than being positive definite in $\Omega \cap \mathbb{R}$. The fact that the kernel has a finite number of negative squares will not imply analyticity. One then resorts to a result of Krein and Langer (see [24, 89], and [20] for the quaternionic case) to obtain a slice hyperholomorphic extension from a given open set.

Chapter 2
Quaternions and Matrices

This chapter contains some basic knowledge on quaternions, Toeplitz and Hankel matrices and we introduce some useful maps which allow to consider, instead of quaternionic matrices, complex matrices of double size. For more information about quaternionic matrices, the interested reader may consult, e.g., Rodman's book [97]. We also recall the notions of Pontryagin and Krein spaces which will be useful in the sequel.

2.1 Quaternions

The set of quaternions, denoted by $\mathbb{H}$, consists of the elements of the form $p = x_0 + x_1\mathbf{i} + x_2\mathbf{j} + x_3\mathbf{k}$, where the three imaginary units $\mathbf{i}, \mathbf{j}, \mathbf{k}$ satisfy

$$\mathbf{i}^2 = \mathbf{j}^2 = \mathbf{k}^2 = -1, \quad \mathbf{ij} = -\mathbf{ji} = \mathbf{k}, \quad \mathbf{ki} = -\mathbf{ik} = \mathbf{j}, \quad \mathbf{jk} = -\mathbf{kj} = \mathbf{i}.$$

The sum and the product of two quaternions

$$p = x_0 + x_1\mathbf{i} + x_2\mathbf{j} + x_3\mathbf{k},$$
$$q = y_0 + y_1\mathbf{i} + y_2\mathbf{j} + y_3\mathbf{k}$$

are defined by

$$p + q = (x_0 + y_0) + (x_1 + y_1)\mathbf{i} + (x_2 + y_2)\mathbf{j} + (x_3 + y_3)\mathbf{k},$$
$$pq = (x_0 y_0 - x_1 y_1 - x_2 y_2 - x_3 y_3) + (x_0 y_1 + x_1 y_0 + x_2 y_3 - x_3 y_2)\mathbf{i} +$$
$$+ (x_0 y_2 - x_1 y_3 + x_2 y_0 + x_3 y_1)\mathbf{j} + (x_0 y_3 + x_1 y_2 - x_2 y_1 + x_3 y_0)\mathbf{k},$$

and with these operations, $\mathbb{H}$ turns out to be a skew field. Given a quaternion p as above, its conjugate is defined to be

D. Alpay et al., *Quaternionic de Branges Spaces and Characteristic Operator Function*, SpringerBriefs in Mathematics, https://doi.org/10.1007/978-3-030-38312-1_2

11

$$\bar{p} = x_0 - x_1\mathbf{i} - x_2\mathbf{j} - x_3\mathbf{k}.$$

The modulus (or norm) of a quaternion is given by the Euclidean norm, i.e.,

$$|p| = \sqrt{p\bar{p}} = \sqrt{\bar{p}p} = \sqrt{x_0^2 + x_1^2 + x_2^2 + x_3^2}.$$

Given $p = x_0 + x_1\mathbf{i} + x_2\mathbf{j} + x_3\mathbf{k}$, its real (or scalar) part x_0 will be denoted also by $\mathrm{Re}(p)$ while $x_1\mathbf{i} + x_2\mathbf{j} + x_3\mathbf{k}$ is the imaginary (or vector) part of p, denoted also by $\mathrm{Im}(p)$.
Let

$$\mathbb{S} = \{p = x_1\mathbf{i} + x_2\mathbf{j} + x_3\mathbf{k} \text{ such that } x_1^2 + x_2^2 + x_3^2 = 1\}$$

be the set of unit purely imaginary quaternions. It is a 2-dimensional sphere in $\mathbb{H}$ identified with $\mathbb{R}^4$. Any element $j \in \mathbb{S}$ satisfies $j^2 = -1$ and thus will be called imaginary unit and the set

$$\mathbb{C}_j = \{z = x + jy, \ x, y \in \mathbb{R}\}$$

is a complex plane.
Given any $p = x + jy \in \mathbb{H}$, we define the 2-sphere $[p]$ associated with it and we have:

$$[p] = \{x + ky \ : \ k \in \mathbb{S}\};$$

we note that if $p \in \mathbb{R}$ then $[p]$ contains only the element p. For the sequel, it is useful to note that the 2-sphere $[p]$ is defined by the second degree equation

$$x^2 - 2\mathrm{Re}(p)x + |p|^2 = 0, \tag{2.1}$$

in fact x is a root of this polynomial if and only if $x \in [p]$.
To any non-real quaternion $p = x_0 + x_1\mathbf{i} + x_2\mathbf{j} + x_3\mathbf{k}$, one associates the imaginary unit j_p defined by

$$j_p = \frac{\mathrm{Im}(p)}{|\mathrm{Im}(p)|}$$

and thus $p \in \mathbb{C}_{j_p}$. It follows that $\mathbb{H} = \cup_{j \in \mathbb{S}} \mathbb{C}_j$.

Assume to fix the imaginary units $i, j, k \in \mathbb{S}$ such that they form a new basis for $\mathbb{H}$. Then, a quaternion p can be written in the form $p = z_1 + z_2 j$ with

$$z_1 = x_0 + ix_1 \text{ and } z_2 = x_2 + ix_3 \ \in \mathbb{C},$$

where we identify $\mathbb{C}$ with the subset $\mathbb{C}_i$ of $\mathbb{H}$ given by the elements of the form $x + iy, x, y \in \mathbb{R}$.
Let $\chi : \mathbb{H} \to \mathbb{C}^{2 \times 2}$ be the map

$$\chi(p) = \begin{pmatrix} z_1 & z_2 \\ -\overline{z_2} & \overline{z_1} \end{pmatrix}. \tag{2.2}$$

Note also that z_1 is determined when $i \in \mathbb{S}$ is fixed, while z_2 is determined once that both $i, j \in \mathbb{S}$ are fixed. Thus the map χ depends on both i, j, has values in $\mathbb{C}_i^{2 \times 2}$ identified with $\mathbb{C}^{2 \times 2}$ thus it should be denoted by $\chi_{i,j}$ (or χ_i to emphasize the values), but for the sake of simplicity we denote it χ. This map allows to translate problems from the quaternionic to the complex matricial setting, since it is an injective homomorphism of rings, i.e.,

$$\chi(p+q) = \chi(p) + \chi(q), \qquad \chi(pq) = \chi(p)\chi(q).$$

The map χ can be extended to matrices in at least two ways. Let $M \in \mathbb{H}^{m \times n}$, $M = (m_{\ell k})$, and write $M = A + Bj$. We can extend the map χ by defining

$$\chi(M) = \begin{pmatrix} A & B \\ -\overline{B} & \overline{A} \end{pmatrix} \quad \text{or} \quad \chi_1(M) = (\chi(m_{\ell k})).$$

Example 2.1.1. *Let* $M = \begin{pmatrix} 0 & i+j \\ k & j \end{pmatrix}$. *Then we have*

$$\chi(M) = \begin{pmatrix} 0 & i & 0 & 1 \\ 0 & 0 & i & 1 \\ 0 & -1 & 0 & -i \\ i & -1 & 0 & 0 \end{pmatrix}, \qquad \chi_1(M) = \begin{pmatrix} 0 & 0 & i & 1 \\ 0 & 0 & -1 & -i \\ 0 & i & 0 & 1 \\ i & 0 & -1 & 0 \end{pmatrix}.$$

Both maps carry the same additive and multiplicative properties. This is well-known for χ, see [110]. We now prove the result for χ_1:

Lemma 2.1.2. *Let M and N be two matrices with quaternionic entries and of compatible sizes. Then*

$$\chi_1(M+N) = \chi_1(M) + \chi_1(N)$$
$$\chi_1(MN) = \chi_1(M)\chi_1(N) \tag{2.3}$$
$$\chi_1(M^*) = \chi_1(M)^*.$$

Proof. To simplify the notation we write the proof for square matrices $M, N \in \mathbb{H}^{u \times u}$. The general case is proved in the same way.
By definition, $\chi_1(M) = (\chi(m_{jk}))_{j,k=1,\dots,u}$ and $\chi_1(N) = (\chi(n_{k\ell}))_{j,k=1,\dots,u}$. Thus

$$(\chi_1(M+N))_{jk} = \chi(m_{jk} + n_{jk})$$
$$= \chi(m_{jk}) + \chi(n_{jk})$$
$$= (\chi_1(M))_{jk} + (\chi_1(N))_{jk}, \quad j, k = 1, \dots u,$$

and

$$(\chi_1(MN))_{j\ell} = \chi((MN)_{j\ell})$$

$$= \chi(\sum_{k=1}^{u}(m_{jk}n_{k\ell}))$$

$$= \sum_{k=1}^{u} \chi(m_{jk})\chi(n_{k\ell})$$

$$= (\chi_1(M)\chi_1(N))_{k\ell}.$$

This proves the first and second equality in (2.3). The proof of last equality is straight-forward and is omitted. □

2.2 Toeplitz and Hankel Matrices

Hermitian non-degenerate Toeplitz and Hankel matrices appear in the finite dimensional theory of $\mathcal{H}(A, B)$ spaces that will be treated in Chapter 6. For this reason, we introduce some basic facts about these matrices.

Recall that a matrix $T \in (\mathbb{H}^{u\times u})^{v\times v}$ is called a block Toeplitz matrix if it is constant on the block diagonals, while it is called a block Hankel matrix if it is constant on the block anti-diagonals. In other words,

$$T = (T_{j-k})_{j,k=1,\dots,v} \quad \text{and} \quad H = (H_{j+k-1})_{j,k=1,\dots,v}$$

where the matrices $T_{-v}, \dots, T_v$ and $H_1, \dots, H_{2v-1}$ belong to $\mathbb{H}^{u\times u}$.

In the previous section, we showed that χ_1 shares some properties of the more classical map χ. The properties in the next result are of key importance and they hold for χ_1 but they are not satisfied by χ.

Lemma 2.2.1. *Assume that $M \in (\mathbb{H}^{u\times u})^{v\times v}$ is a block Toeplitz (resp. block Hankel) matrix. Then the $\mathbb{C}^{2uv\times 2uv}$ matrix $\chi_1(M)$ is block Toeplitz (resp. block Hankel). Let $Z_{uv} \in \mathbb{H}^{uv\times uv}$ be defined by*

$$Z_{uv} = \begin{pmatrix} 0 & I_u & 0 & \dots & 0 & 0 \\ 0 & 0 & I_u & 0 & \dots & 0 \\ & & \ddots & & & \\ 0 & 0 & 0 & \dots & I_u & 0 \\ 0 & 0 & 0 & \dots & 0 & 0 \end{pmatrix}.$$

Then

$$\chi_1(Z_{uv}) = \begin{pmatrix} 0 & I_{2u} & 0 & \cdots & 0 & 0 \\ 0 & 0 & I_{2u} & 0 & \cdots & 0 \\ & & & \ddots & & \\ 0 & 0 & 0 & \cdots & I_{2u} & 0 \\ 0 & 0 & 0 & \cdots & 0 & 0 \end{pmatrix} = Z_{2uv}. \qquad (2.4)$$

Proof. The proof easily follows using standard arguments.

Let $T = (T_{j-k})_{j,k=0}^{N}$ be a Hermitian Toeplitz block matrix with blocks in $\mathbb{H}^{u \times u}$. Then

$$T - ZTZ^* = C^* J_0 C \qquad (2.5)$$

where

$$J_0 = \begin{pmatrix} I_u & 0 \\ 0 & -I_u \end{pmatrix} \quad \text{and} \quad C = \begin{pmatrix} I_u & 0 & \cdots & 0 \\ \frac{T_0}{2} & T_1 & \cdots & T_{N-1} \end{pmatrix}.$$

Note that (2.5) is a special case of the Stein equation

$$P - A^* P A = C^* J C,$$

where J is any signature matrix (see also (6.12)) and note also that $C^* J C$ expresses the displacement rank of T with respect to Z. The reader interested in more information on displacement ranks and structured matrices may consult [44, 61, 83, 84, 105, 106], and [28, 29, 31, 32, 68] for a study of these topics using reproducing kernel methods.

The Gohberg-Heinig formula to invert block Toeplitz matrices holds for general algebras (see [79, 80]) and, in particular, in the present setting. Since the map χ_1 keeps the underlying algebraic structure, see Lemma 2.2.1, we can work in the complex setting. Let $A = (a_{j-k})_{j,k=1,\dots,n}$ be a Toeplitz matrix. Then the first step to invert A is to consider, for $j = 0, \dots, n$ the equations

$$\sum_{k=0}^{N} a_{j-k} x_k = \delta_{0j} I_u \qquad (2.6)$$

$$\sum_{k=0}^{N} a_{k-j} z_{-k} = \delta_{0j} I_u \qquad (2.7)$$

$$\sum_{k=0}^{N} w_k a_{j-k} = \delta_{0j} I_u \qquad (2.8)$$

$$\sum_{k=0}^{N} y_{-k} a_{k-j} = \delta_{0j} I_u \qquad (2.9)$$

where the unknowns are in $\mathbb{H}^{u \times u}$. Assuming that they are solvable, one has $x_0 = y_0$ and $z_0 = w_0$. Note that (x_k) is the first block column of A^{-1} while, by setting

$$
S = \begin{pmatrix}
0 & 0 & \cdots & 0 & I_u \\
0 & 0 & \cdots & I_u & 0 \\
0 & 0 & \ddots & 0 & 0 \\
0 & I_u & \cdots & & \\
I_u & 0 & \cdots & 0 & 0
\end{pmatrix},
\tag{2.10}
$$

(z_{-k}) is the first block column of $SA^{-1}S$. Similarly (w_k) is the first block row of A^{-1} while (y_{-k}) is the first block column of $SA^{-1}S$.

As we mentioned, the Gohberg-Heinig formula in [80] gives the inverse of a block Toeplitz matrix T:

Theorem 2.2.2. *Assume that the equations* (2.6)–(2.9) *are solvable, and that x_0 or z_0 is invertible. Then the other is also invertible, T is invertible and one has the formula*

$$
T^{-1} = \begin{pmatrix}
x_0 & 0 & \cdots & 0 \\
x_1 & x_0 & \cdots & 0 \\
\vdots & \vdots & \ddots & \vdots \\
x_N & x_{N-1} & \cdots & x_0
\end{pmatrix} x_0^{-1} \begin{pmatrix}
y_0 & y_{-1} & \cdots & y_{-N} \\
0 & y_0 & \cdots & y_{1-N} \\
\vdots & \vdots & \ddots & \vdots \\
0 & 0 & \cdots & y_0
\end{pmatrix} -
$$

$$
- \begin{pmatrix}
0 & 0 & \cdots & 0 \\
z_{-N} & \cdots & 0 & 0 \\
\vdots & \ddots & \vdots & \vdots \\
z_{-1} & \cdots & z_{-N} & 0
\end{pmatrix} z_0^{-1} \begin{pmatrix}
0 & w_N & \cdots & w_1 \\
\vdots & \vdots & \ddots & \vdots \\
0 & 0 & \cdots & w_N \\
0 & 0 & \cdots & 0
\end{pmatrix}.
\tag{2.11}
$$

Moreover, we have:

Theorem 2.2.3. *Let $T = (T_{j-k})_{j,k=1,\ldots,N}$ and assume that T_N is invertible. We obtain*

$$
T^{-1} - Z_u^* T^{-1} Z_u = \begin{pmatrix} x_0 \\ x_1 \\ \vdots \\ x_N \end{pmatrix} x_0^{-1} \begin{pmatrix} x_0 \\ x_1 \\ \vdots \\ x_N \end{pmatrix}^* - \begin{pmatrix} 0 \\ z_0 \\ \vdots \\ z_N \end{pmatrix} z_0^{-1} \begin{pmatrix} 0 \\ z_0 \\ \vdots \\ z_N \end{pmatrix}^*
\tag{2.12}
$$

and the Christoffel-Darboux formula

$$
\sum_{a,b=0}^{N} x^a y^b (T^{-1})_{ab} = \frac{Q_N(x) x_0^{-1} Q_N(y)^* - xy P_N(x) z_0^{-1} P_N(y)^*}{1 - xy}
\tag{2.13}
$$

where

$$P_N(x) = \gamma_{0N}^{(N)} + x\gamma_{1N}^{(N)} + \cdots + x^N \gamma_{NN}^{(N)} \qquad (2.14)$$

and

$$Q_N(x) = \gamma_{00}^{(N)} + x\gamma_{10}^{(N)} + \cdots + x^N \gamma_{N0}^{(N)},$$

where $(\gamma_{ab}^{(N)})_{a,b=0}^N$ is the block entry decomposition of T_N^{-1} into $\mathbb{H}^{u\times u}$ blocks.

Proof. The proof is a consequence of Theorem 2.2.2 and mimics the one in the complex case, see [33, Section 4]. □

Remark 2.2.4. Endow the space of matrix-valued polynomials of the form

$$p(z) = \sum_{a=0}^N p_a z^a, \quad p_0, \ldots, p_N \in \mathbb{H}^{u\times u}$$

with the possibly indefinite inner product

$$\langle p, q \rangle = \left(q_0^* \ q_1^* \cdots q_N^* \right) T_N \begin{pmatrix} p_0 \\ p_1 \\ \vdots \\ p_N \end{pmatrix}, \quad q(z) = \sum_{a=0}^N q_a z^a, \quad q_0, \ldots, q_N \in \mathbb{H}^{u\times u}.$$

Recalling the definition of P_N in (2.14), we see that

$$\langle P_N, q \rangle = 0$$

for every polynomial $q(z) = \sum_{a=0}^N q_a z^a$ for which $q_N = 0$ since the coefficients of P_N are the entries of the last (block) column of T_N^{-1}. This justifies the terminology *orthogonal polynomial* for P_N, even in the nonpositive case. In the positive case, the entries of T_N are moments of a positive measure, and one gets back the classical definition of orthogonal polynomials.

It is interesting to note that, in the complex case, the distribution of zeros of P_N with respect to the unit circle is characterized by a theorem of Krein and it is given in terms of the signature of T_N; see [88]. It has been extended to the matrix-valued case in the papers [33, 82].

Krein's theorem is a particular case of the following result, when T is assumed to be a Toeplitz matrix. See [67, 71] for more general results. We point out that the quaternionic counterpart of this result is given in Theorem 6.5.1.

Theorem 2.2.5. *Let $T \in \mathbb{C}^{n\times n}$ be an invertible Hermitian matrix, with $\nu \geq 0$ negative eigenvalues. Assume furthermore that*

$$\left(1 \ z \ldots z^n \right) T^{-1} \left(1 \ \overline{w} \ldots \overline{w}^n \right)^t = \frac{A(z)\overline{A(w)} - z\overline{w}B(z)\overline{B(w)}}{1 - z\overline{w}} \qquad (2.15)$$

where A and B are polynomials of degree n. Then, A has ν zeros inside $\mathbb{D}$ and B has ν zeros outside $\mathbb{D}$. They have no zeros on the unit circle.

Proof. We split the proof in several steps.

STEP 1: *The number of negative squares of the kernel*

$$\left(1\ z\ \dots\ z^n\right) T^{-1} \left(1\ \overline{w}\ \dots\ \overline{w}^n\right)^t$$

is equal to the number of negative eigenvalues of T.
This follows from the spectral theorem for Hermitian matrices.

STEP 2: *A and B have no common zeros.*

Indeed, a common zero, say z_0, will be such that

$$\left(1\ z_0\ \cdots\ z_0^n\right) T^{-1} \left(1\ \overline{w}\ \cdots\ \overline{w}^n\right)^t = 0, \quad \forall w \in \mathbb{C},$$

and hence we would have

$$\left(1\ z_0\ \cdots\ z_0^n\right) T^{-1} = \left(0\ 0\ \cdots\ 0\right),$$

contradicting the invertibility of T.

STEP 3: *The polynomials A and B have no zeros on the unit circle.*
Assume by contradiction that $A(z_0) = 0$, with $|z_0| = 1$ (the same argument would work for B). Rewriting formula (2.15) as

$$(1 - z\overline{w}) \left(1\ z\ \dots\ z^n\right) T^{-1} \left(1\ \overline{w}\ \dots\ \overline{w}^n\right)^t = A(z)\overline{A(w)} - z\overline{w}B(z)\overline{B(w)} \quad (2.16)$$

and setting $z = w = z_0$ implies that $B(z_0) = 0$, which cannot be by the previous step.

STEP 4: *The function $A^{-1}B$ is a generalized Schur function and can be written as*

$$A^{-1}B = S_1 S_2^{-1}$$

where S_1 is a Blaschke product of degree $n - \nu$ and S_2 is a Blaschke product of degree ν, without common zeros.

The fact that $A^{-1}B$ is a generalized Schur functions follows from the definition. The second part of the assertion follows from elementary facts on rational functions, but it follows also from the Krein-Langer's theorem.

STEP 5: *We conclude the proof.*

We set

$$S_1(z) = \prod_{k=1}^{n-\nu} \frac{z - w_k}{1 - z\overline{w}_k} \quad \text{and} \quad S_2(z) = \prod_{j=1}^{\nu} \frac{z - v_j}{1 - z\overline{v}_j}.$$

Write $A(z)S_1(z) = B(z)S_2(z)$, i.e.,

$$A(z) \prod_{k=1}^{n-\nu} \frac{z - w_k}{1 - z\overline{w_k}} = B(z) \prod_{j=1}^{\nu} \frac{z - v_j}{1 - z\overline{v_j}}$$

or

$$A(z) \prod_{j=1}^{\nu} (1 - z\overline{v_j}) \prod_{k=1}^{n-\nu} (z - w_k) = z B(z) \prod_{j=1}^{\nu} (z - v_j) \prod_{k=1}^{n-\nu} (1 - z\overline{w_k}),$$

and assume that $A(w) = 0$ with $w \in \mathbb{D}$. We note that the v_j's and the w_k's may be repeated. Since A and B have no common zeros and $A(0) \neq 0$, we have $S_2(w) = 0$, and A has ν zeros inside $\mathbb{D}$. The other $n - \nu$ zeros correspond to the poles of S_1. Suppose now that $S_1(\nu) = 0$. Since S_1 and S_2 have no common zeros we have that $B(\nu) = 0$. So B has $n - \nu$ zeros inside $\mathbb{D}$. The other ν zeros correspond to the poles of S_1. □

Hermitian Hankel matrices are automatically real-valued; so the case of interest is that of block Hankel matrices $H = (H_{ab})_{a,b=0}^N$, where each $H_{j+k} \in \mathbb{H}^{u \times u}$ is self-adjoint. If S is as in (2.10) then,

$$T = HS = (H_{N+|a-b|})_{a,b=0}^N \tag{2.17}$$

is a (non-Hermitian) Toeplitz matrix, to which the Gohberg-Heinig formula is applicable.
So, it follows from (2.17) that

$$
H^{-1} = \begin{pmatrix} x_n & x_{n-1} & \cdots & x_0 \\ x_{n-1} & \cdots & x_0 & 0 \\ \vdots & \vdots & \ddots & \vdots \\ x_0 & 0 & \cdots & 0 \end{pmatrix} x_0^{-1} \begin{pmatrix} y_0 & y_{-1} & \cdots & y_{-n} \\ 0 & y_0 & \cdots & y_{1-n} \\ \vdots & \vdots & \ddots & \vdots \\ 0 & 0 & \cdots & y_0 \end{pmatrix} -
$$
$$
- \begin{pmatrix} z_{-1} & \cdots & z_{-n} & 0 \\ \vdots & \ddots & \vdots & \vdots \\ z_{-n} & 0 & \cdots & 0 \\ 0 & 0 & \cdots & 0 \end{pmatrix} z_0^{-1} \begin{pmatrix} 0 & w_n & \cdots & w_1 \\ \vdots & \vdots & \ddots & \vdots \\ 0 & 0 & \cdots & w_n \\ 0 & 0 & \cdots & 0 \end{pmatrix}. \tag{2.18}
$$

Similarly to Theorem 2.2.5, one has the following result which holds in particular for Hankel matrices.

Theorem 2.2.6. *Let $T \in \mathbb{C}^{n \times n}$ be an invertible Hermitian matrix, with $\nu \geq 0$ negative eigenvalues. Assume furthermore that*

$$\left(1 \ z \ldots z^n\right) T^{-1} \left(1 \ \overline{w} \ldots \overline{w}^n\right)^t = \frac{A(z)\overline{A(w)} - B(z)\overline{B(w)}}{z + \overline{w}} \tag{2.19}$$

where A and B are polynomials of degree n. Then, A has ν zeros inside $\mathbb{C}_r$ and B has ν zeros outside the half-plane $\mathbb{C}_r$. They have no zeros on the imaginary axis.

2.3   Some Remarks in Functional Analysis

To develop our analysis, we will need the quaternionic counterpart of some classical results from functional analysis; these can be found in the works [19, 21, 23]. In this section we only briefly recall some of the definitions, to set the framework and the notation. Notions and results related to the spectrum of an operator in a quaternionic vector space involve the notion of slice hyperholomorphic functions and are postponed to the next section. Some of our results are in the Pontryagin space setting, so we will recall below the notion of quaternionic Pontryagin space and quaternionic reproducing kernel Pontryagin space.

Let $\mathcal{V}$ be a right quaternionic vector space, namely, a linear space over $\mathbb{H}$ where the scalars are multiplied on the right. A map $T : \mathcal{V} \to \mathcal{V}$ is said to be a right linear operator if

$$T(u + v) = T(u) + T(v), \quad T(up) = T(u)p, \text{ for all } p \in \mathbb{H}, \ u, v \in \mathcal{V}.$$

An analogous definition can be given for a left quaternionic vector space and a left linear operator.

Definition 2.3.1. *Let $\mathcal{V}$ be a right quaternionic vector space. The $\mathbb{H}$-valued map $[\cdot, \cdot] : \mathcal{V} \times \mathcal{V} \to \mathbb{H}$ is called a Hermitian form if it satisfies the following conditions for all $u, v, w \in \mathcal{V}$ and $p, q \in \mathbb{H}$:*

$$[u, v + w] = [u, v] + [u, w], \tag{2.20}$$
$$[u, v] = \overline{[v, u]}, \tag{2.21}$$
$$[up, vq] = \overline{q}[u, v]p. \tag{2.22}$$

When one endows $\mathcal{V}$ with a two-sided quaternionic linear structure one requires moreover that

$$[pu, v] = [u, \overline{p}v]. \tag{2.23}$$

We will call such a form a (possibly degenerate and nonpositive) inner product. If such a form is positive and

$$\|u\| = \sqrt{[u, u]}$$

defines a norm such that $\mathcal{V}$ is complete, $\mathcal{V}$ is said a (quaternionic) Hilbert space.

Definition 2.3.2. *Let $\mathcal{V}$ be a right quaternionic vector space and let $[\cdot, \cdot]$ be an associated Hermitian form. The pair $(\mathcal{V}, [\cdot, \cdot])$ is called a (quaternionic) Krein space if it can be written as a direct and orthogonal sum (fundamental decomposition)*

$$\mathcal{V} = \mathcal{V}_+ \dotplus \mathcal{V}_-,$$

where $(\mathcal{V}_+, [\cdot, \cdot])$ *and* $(\mathcal{V}_-, -[\cdot, \cdot])$ *are right quaternionic Hilbert spaces. It is called a (quaternionic) Pontryagin space if* $\mathcal{V}_-$ *is finite dimensional in one (and hence all) fundamental decompositions.*

Functional analysis in Pontryagin spaces can be seen as a "finite dimensional perturbation" of functional analysis in Hilbert spaces. There are results which do not extend to Krein spaces. For instance, an important result which does not extend is that the adjoint of a contraction between two quaternionic Pontryagin spaces of same index is still a contraction, see [23, Theorem 5.7.10].

For the sake of completeness, we recall the definition of a reproducing kernel Pontryagin space.

Definition 2.3.3. *The Pontryagin space* $(\mathcal{P}, [\cdot, \cdot])$ *of* $\mathbb{H}^n$*-valued functions defined on some set* Ω *is called a reproducing kernel Pontryagin space if there exists a* $\mathbb{H}^{n \times n}$*-valued function* $K(a, b)$, *called the reproducing kernel, and with the following properties:*
(1) The function $K_b h : a \mapsto K(a, b)h$ *belongs to* $\mathcal{P}$ *for every choice of* $b \in \Omega$ *and* $h \in \mathbb{H}^n$.
(b) With b and h as above, one has

$$h^* f(b) = [f, K_b h] \tag{2.24}$$

for every $f \in \mathcal{P}$.

An example of a finite dimensional reproducing kernel Pontryagin space can be built from the kernel in (2.13), in fact that kernel has a finite number of negative squares.

Remark 2.3.4. We note that there is a one-to-one correspondence between quaternionic reproducing kernel Pontryagin spaces and Hermitian kernels which can be written as difference of two positive kernels, one being of finite rank.

We also note that Definition 2.3.3 extends to Krein spaces. A necessary and sufficient condition for a function to be the reproducing kernel of a quaternionic reproducing kernel Krein space is that it can be written as a difference of two positive definite functions. The associated reproducing kernel Krein space will not be unique in general; see [107, Section 13] and [3] for a discussion in the complex setting.

We note that the quaternionic versions of the closed-graph theorem and the inverse mapping theorem (see, e.g., the proof of Proposition 10.3.2) will be needed later in this work. For these and other results we refer the reader to [23].

Chapter 3
Slice Hyperholomorphic Functions

Functions of a quaternionic variable with properties generalizing holomorphicity from the complex to the quaternionic setting can be defined in various different ways. The notion of hyperholomorphicity which looks more suitable for some applications to operator theory is the so-called slice hyperholomorphicity; this is the main object of this chapter. This function theory has been evolving rapidly and for more information the reader may consult the books [23, 58, 60, 72, 76] and the references therein. This chapter also contains some basic notions on the so-called S-functional calculus which is nowadays very well developed. Dozens of papers have been written on this topic, in various directions, and here with no sake of completeness we mention [12–14, 16, 18, 22, 52, 54–57, 59, 74]. For more details, we refer to [23, 53, 60] and the references therein.

3.1 Slice Hyperholomorphic Functions

In the development of slice hyperholomorphic functions we follow our books [23] and [60] but for the scalar-valued case see also [76]. There are different ways of defining slice hyperholomorphic functions. In this section we repeat only the basic notions that we will use in the sequel, and for more information we refer the reader to the aforementioned books. Below we follow the definition related with the Fueter mapping theorem (which is a construction giving Fueter regular functions starting from holomorphic functions), which is also a simplified version of the approach used in [77, 78] where the authors make use of the so-called stem functions.

Definition 3.1.1. *Let $U \subseteq \mathbb{H}$. We say that U is* axially symmetric *if $[q] \subset U$ for any $q \in U$.*
We say that U is a slice domain *(s-domain for short) if $U \cap \mathbb{R}$ is non-empty and if $U \cap \mathbb{C}_j$ is a domain in $\mathbb{C}_j$ for all $j \in \mathbb{S}$.*

© The Author(s), under exclusive license to Springer Nature Switzerland AG 2020
D. Alpay et al., *Quaternionic de Branges Spaces and Characteristic Operator Function*, SpringerBriefs in Mathematics,
https://doi.org/10.1007/978-3-030-38312-1_3

Definition 3.1.2 (Slice functions). *Let $U \subseteq \mathbb{H}$ be an axially symmetric open set and let $\mathcal{U} = \{(u, v) \in \mathbb{R}^2 : u + jv \in U$ for some $j \in \mathbb{S}\}$. A function $f : U \to \mathbb{H}$ is called a left slice function, if it is of the form*

$$f(q) = f_0(u, v) + jf_1(u, v) \quad for\ q = u + jv \in U$$

where two functions $f_0, f_1 : \mathcal{U} \to \mathbb{H}$ satisfy the compatibility conditions

$$f_0(u, -v) = f_0(u, v), \qquad f_1(u, -v) = -f_1(u, v). \tag{3.1}$$

A function $f : U \to \mathbb{H}$ is called a right slice function if it is of the form

$$f(q) = f_0(u, v) + f_1(u, v)j \quad for\ q = u + jv \in U$$

where two functions $f_0, f_1 : \mathcal{U} \to \mathbb{H}$ satisfy (3.1).
If f is a left (or right) slice function such that f_0 and f_1 are real-valued, then f is called intrinsic.

Definition 3.1.3 (Slice hyperholomorphic functions). *Let $U \subseteq \mathbb{H}$ be an axially symmetric open set and let $\mathcal{U} = \{(u, v) \in \mathbb{R}^2 : u + jv \in U$ for some $j \in \mathbb{S}\}$. Let $f : U \to \mathbb{H}$ be a left slice function*

$$f(q) = f_0(u, v) + jf_1(u, v) \quad for\ q = u + jv \in U.$$

If f_0 and f_1 satisfy the Cauchy-Riemann-equations

$$\frac{\partial}{\partial u} f_0(u, v) - \frac{\partial}{\partial v} f_1(u, v) = 0 \tag{3.2}$$

$$\frac{\partial}{\partial v} f_0(u, v) + \frac{\partial}{\partial u} f_1(u, v) = 0, \tag{3.3}$$

then f is called left slice hyperholomorphic. If f is a right slice function

$$f(q) = f_0(u, v) + f_1(u, v)j \quad for\ q = u + jv \in U$$

and f_0 and f_1 satisfy the Cauchy-Riemann-equation (3.2), then f is called right slice hyperholomorphic.
We denote the sets of left and right slice hyperholomorphic functions on U by $\mathcal{SH}_L(U)$ and $\mathcal{SH}_R(U)$, respectively. The set of intrinsic slice hyperholomorphic functions on U will be denoted by $\mathcal{N}(U)$.

When we will write "slice hyperholomorphic" function, we will mean "left slice hyperholomorphic" function. A fundamental property of slice functions, in particular slice hyperholomorphic, is the following structure formula.

Theorem 3.1.4 (The Structure formula (or Representation formula)). *Let $U \subset \mathbb{H}$ be axially symmetric and let $i \in \mathbb{S}$. A function $f : U \to \mathbb{H}$ is a left slice function on U if and only if for any $q = u + jv \in U$*

$$f(q) = \frac{1}{2}\Big[f(\bar{z}) + f(z)\Big] + \frac{1}{2}ji\Big[f(\bar{z}) - f(z)\Big] \tag{3.4}$$

with $z = u + iv$. A function $f : U \to \mathbb{H}$ is a right slice function on U if and only if for any $q = u + jv \in U$

$$f(q) = \frac{1}{2}\Big[f(\bar{z}) + f(z)\Big] + \frac{1}{2}\Big[f(\bar{z}) - f(z)\Big]ij \tag{3.5}$$

with $z = u + iv$.

The pointwise multiplication of two slice (hyperholomorphic) functions is not, in general, a function of the same type. Thus we need a suitable notion of multiplication:

Definition 3.1.5. *Let $U \subset \mathbb{H}$ be an axially symmetric open set. If f, g are slice functions in U with $f(q) = f_0 + jf_1$ and $g = g_0 + jg_1$ for $q = u + jv \in U$, we define their left slice product as*

$$f \star_l g := f_0 g_0 - f_1 g_1 + j (f_0 g_1 + f_1 g_0). \tag{3.6}$$

In particular, the $\star_l$-product, in short $\star$-product, is defined for $f, g \in \mathcal{SH}_L(U)$ and $f \star_l g \in \mathcal{SH}_L(U)$.
If f, g are right slice functions with $f(q) = f_0(u, v) + f_1(u, v)j$ and $g(q) = g_0(u, v) + g_1(u, v)j$ for $q = u + jv \in U$, we define their right slice product as

$$f \star_r g := f_0 g_0 - f_1 g_1 + (f_0 g_1 + f_1 g_0) j. \tag{3.7}$$

In particular, the $\star_r$-product is defined for $f, g \in \mathcal{SH}_R(U)$ and $f \star_r g \in \mathcal{SH}_R(U)$.

Remark 3.1.6. We note that if $f \in \mathcal{N}(U)$ and $g \in \mathcal{SH}_L(U)$, then $f \star_l g = g \star_l f = fg \in \mathcal{SH}_L(U)$ and similarly when $g \in \mathcal{SH}_R(U)$.

We are also in need of the two definitions below:

Definition 3.1.7. *If $f \in \mathcal{SH}_L(U)$ with $f(q) = f_0 + jf_1$ we define*

- *the conjugate $f^c = \bar{f_0} + j\bar{f_1}$,*
- *the symmetrization $f^s = f \star_l f^c = f^c \star_l f$, which is given by*

$$f^s = |f_0|^2 - |f_1|^2 + 2j\mathrm{Re}(f_1 \bar{f_2}).$$

If $f \in \mathcal{SH}_R(U)$ with $f(q) = f_0 + f_1 j$ we define

- the conjugate $f^c = \overline{f_0} + \overline{f_1}j$,
- the symmetrization $f^s = f \star_r f^c = f^c \star_r f$, which is given by the formula above.

Given a left or right slice hyperholomorphic function, we may ask if there is an inverse with respect to the $\star$-product. The answer is positive as is given in the following result:

Proposition 3.1.8. *The slice hyperholomorphic inverse is given in the following result.*

- *Let $f \in \mathcal{SH}_L(U)$ be non identically zero, then the left slice hyperholomorphic inverse $f^{-\star_l}$ given by*

$$f^{-\star_l} = (f^s)^{-1} \star_l f^c = (f^s)^{-1} f^c$$

is defined on $U \setminus \{q \in U : f(q) = 0\}$ and satisfies $f^{-\star_l} \star_l f = f \star_l f^{-\star_l} = 1$.
- *Let $f \in \mathcal{SH}_R(U)$ be non identically zero, then the right slice hyperholomorphic inverse $f^{-\star_r}$ given by*

$$f^{-\star_r} = f^c \star_r (f^s)^{-1} = f^c (f^s)^{-1}$$

is defined on $U \setminus \{q \in U : f(q) = 0\}$ and satisfies $f^{-\star_r} \star_r f = f \star_r f^{-\star_r} = 1$.
- *When $f \in \mathcal{N}(U)$ is non identically zero, then $f^{-\star_r} = f^{-\star_l} = f^{-1}$.*

As a consequence of the Structure formula and the Residue Theorem, one can prove the Cauchy formulas with slice hyperholomorphic Cauchy kernels. These kernels are defined outside a 2-sphere, see (2.1).

Definition 3.1.9 (Slice hyperholomorphic Cauchy kernels). *Let $q, s \in \mathbb{H}$ with $q \notin [s]$.*

- *The left Cauchy kernel $S_L^{-1}(s, q)$ is defined as*

$$S_L^{-1}(s, q) := -(q^2 - 2\mathrm{Re}(s)q + |s|^2)^{-1}(q - \bar{s}).$$

- *The right Cauchy kernel $S_R^{-1}(s, q)$ is defined as*

$$S_R^{-1}(s, q) := -(q - \bar{s})(q^2 - 2\mathrm{Re}(s)q + |s|^2)^{-1}.$$

We recall that a slice Cauchy domain is an axially symmetric open set $U \subset \mathbb{H}$ such that $U \cap \mathbb{C}_j$ is a Cauchy domain in $\mathbb{C}_j$ for all $j \in \mathbb{S}$; in other words, the boundary of $U \cap \mathbb{C}_j$ is the union of a finite number of non intersecting piecewise continuously differentiable Jordan curves in $\mathbb{C}_j$.

Theorem 3.1.10 (The Cauchy formulas). *Let $U \subset \mathbb{H}$ be a bounded slice Cauchy domain, let $j \in \mathbb{S}$ and set $ds_j = ds(-j)$. If f is a (left) slice hyperholomorphic function on a set that contains $\overline{U}$ then*

$$f(q) = \frac{1}{2\pi} \int_{\partial(U \cap \mathbb{C}_j)} S_L^{-1}(s, q) \, ds_j \, f(s), \qquad \text{for any } q \in U. \qquad (3.8)$$

If f is a right slice hyperholomorphic function on a set that contains $\overline{U}$, then

$$f(q) = \frac{1}{2\pi} \int_{\partial(U \cap \mathbb{C}_j)} f(s) \, ds_j \, S_R^{-1}(s, q), \qquad \text{for any } q \in U. \qquad (3.9)$$

These integrals depend neither on U nor on the imaginary unit $j \in \mathbb{S}$.

3.2   The S-Resolvent Operators and the S-Spectrum

Slice hyperholomorphic functions can be defined also for vector-valued functions, see [23] and references therein. As it happens for the class of holomorphic functions there is the concept of strong and weakly slice hyperholomorphicity. Here we just need a readaptation of the scalar-valued case previously introduced.

Definition 3.2.1 (Slice hyperholomorphic functions vector-valued). *Let $U \subseteq \mathbb{H}$ be an axially symmetric open set and let*

$$\mathcal{U} = \{(u, v) \in \mathbb{R}^2 : u + jv \in U, \ j \in \mathbb{S}\}.$$

A function $f : U \to \mathcal{X}_L$ with values in a quaternionic left Banach space $\mathcal{X}_L$ is called a left slice function, if is of the form

$$f(q) = f_0(u, v) + j f_1(u, v) \qquad \text{for } q = u + jv \in U$$

with two functions $f_0, f_1 : \mathcal{U} \to \mathcal{X}_L$ that satisfy the compatibility condition (3.1). If in addition f_0 and f_1 satisfy the Cauchy-Riemann-equations (3.2), then f is called strongly left slice hyperholomorphic.
A function $f : U \to \mathcal{X}_R$ with values in a quaternionic right Banach space $\mathcal{X}_R$ is called a right slice function if it is of the form

$$f(q) = f_0(u, v) + f_1(u, v) j \qquad \text{for } q = u + jv \in U$$

with two functions $f_0, f_1 : \mathcal{U} \to \mathcal{X}_R$ that satisfy the compatibility condition (3.1). If in addition f_0 and f_1 satisfy the Cauchy-Riemann-equations (3.2), then f is called strongly right slice hyperholomorphic.

Functions with values in a quaternionic Banach algebra can be multiplied, adapting the definition in the scalar-valued case:

Definition 3.2.2. *Let $U \subset \mathbb{H}$ be an axially symmetric open set and let $\mathcal{X}$ be a two-sided quaternionic Banach algebra. For two functions $f, g \in \mathcal{SH}_L(U, \mathcal{X})$ with*

$f(q) = f_0 + jf_1$ and $g = g_0 + jg_1$ for $q = u + jv \in U$, we define their left slice hyperholomorphic product as

$$f \star_l g := f_0 g_0 - f_1 g_1 + j (f_0 g_1 + f_1 g_0). \qquad (3.10)$$

For two functions $f, g \in \mathcal{SH}_R(U, \mathcal{X})$ with $f(q) = f_0(u, v) + f_1(u, v)j$ and $g(q) = g_0(u, v) + g_1(u, v)j$ for $q = u + jv \in U$, we define their right slice hyperholomorphic product as

$$f \star_r g := f_0 g_0 - f_1 g_1 + (f_0 g_1 + f_1 g_0) j. \qquad (3.11)$$

Remark 3.2.3. It is immediate that the $\star_l$-product of two left slice hyperholomorphic functions is again left slice hyperholomorphic and that the $\star_r$-product of two right slice hyperholomorphic functions is again right slice hyperholomorphic. If moreover $U = B_r(0)$, then f, g admit power series expansions. More precisely, if f and g are left slice hyperholomorphic with $f(q) = \sum_{n=0}^{+\infty} q^n a_n$ and $g(q) = \sum_{n=0}^{+\infty} q^n b_n$ with $a_n, b_n \in \mathcal{X}$, then

$$(f \star_l g)(q) := \sum_{n=0}^{+\infty} q^n \left(\sum_{\ell=0}^{n} a_\ell b_{n-\ell} \right).$$

Similarly, if f and g are right slice hyperholomorphic with $f(q) = \sum_{n=0}^{+\infty} a_n q^n$ and $g(q) = \sum_{n=0}^{+\infty} b_n q^n$ with $a_n, b_n \in \mathcal{X}$, then

$$(f \star_r g)(q) := \sum_{n=0}^{+\infty} \left(\sum_{\ell=0}^{n} a_\ell b_{n-\ell} \right) q^n.$$

In the sequel, we will need the following definition:

Definition 3.2.4. Let U be an axially symmetric s-domain in $\mathbb{H}$. We say that a function $f : U \to \mathbb{H}$ is slice hypermeromorphic in U if f is slice hyperholomorphic in $U' \subset U$ such that $(U \setminus U') \cap \mathbb{C}_j$ has no accumulation point in $U \cap \mathbb{C}_j$ for $j \in \mathbb{S}$, and every point in $U \setminus U'$ is a pole.

More in general, in the case of functions with values in a quaternionic Banach space, we have

Definition 3.2.5. Let $\mathcal{X}$ be a two-sided quaternionic Banach space. We say that a function $f : U \to \mathcal{X}$ is (weakly) slice hypermeromorphic if for any $\Lambda \in \mathcal{X}^*$ the function $\Lambda f : U \to \mathbb{H}$ is slice hypermeromorphic in U.

The crucial notions in quaternionic operator theory are those of S-spectrum and of S-resolvent set which replace the classical concept of spectrum and resolvent set.

Definition 3.2.6. Let $\mathcal{X}$ be a two-sided quaternionic Banach space. Let $T \in \mathcal{B}(\mathcal{X})$. For $s \in \mathbb{H}$, we set

$$Q_s(T) := T^2 - 2\text{Re}(s)T + |s|^2 I.$$

The S-resolvent set $\rho_S(T)$ *of* T *is*

$$\rho_S(T) := \{s \in \mathbb{H} : \mathcal{Q}_s(T) \text{ is invertible in } \mathcal{B}(\mathcal{X})\},$$

while the S-spectrum $\sigma_S(T)$ *of* T *is*

$$\sigma_S(T) := \mathbb{H} \setminus \rho_S(T).$$

For $s \in \rho_S(T)$, *the operator* $\mathcal{Q}_s(T)^{-1} \in \mathcal{B}(\mathcal{X})$ *is called the* pseudo-resolvent *of* T *at* s.

Observe that if $T \in \mathcal{B}(\mathcal{X})$ then the sets $\rho_S(T)$ and $\sigma_S(T)$ are axially symmetric, in fact we have, see [60]:

Theorem 3.2.7 *(Structure of the S-spectrum). Let* T *be a linear operator acting on a quaternionic linear space and let* $p = p_0 + jp_1 \in \sigma_S(T)$. *Then all the elements of the sphere* $[p_0 + jp_1]$ *belong to* $\sigma_S(T)$.

Moreover, we can give the definition of S-spectral radius, see [60]:

Definition 3.2.8. *Let* $\mathcal{H}$ *be a quaternionic Hilbert space and* $T \in \mathcal{B}(\mathcal{H})$. *We call* S-spectral radius *of* T *the nonnegative real number*

$$r_S(T) := \sup\{|s| : s \in \sigma_S(T)\}.$$

We have:

Theorem 3.2.9 (The S-spectral radius of T). *Let* $\mathcal{H}$ *be a quaternionic Hilbert space,* $T \in \mathcal{B}(\mathcal{H})$, *and let* $r_S(T)$ *be its* S-spectral radius. Then

$$r_S(T) = \lim_{n \to \infty} \|T^n\|^{1/n}. \tag{3.12}$$

As a consequence of the fact that for slice hyperholomorphic functions there are two different Cauchy kernels, the functional calculus based on slice hyperholomorphicity has two resolvent operators.

Definition 3.2.10. *Let* $T \in \mathcal{B}(\mathcal{X})$. *For* $s \in \rho_S(T)$, *we define the* left S-resolvent operator *as*

$$S_L^{-1}(s, T) = -\mathcal{Q}_s(T)^{-1}(T - \bar{s}I),$$

and the right S-resolvent operator *as*

$$S_R^{-1}(s, T) = -(T - \bar{s}I)\mathcal{Q}_s(T)^{-1}.$$

The noncommutative setting shows other peculiarities, for example the fact that the S-resolvent equation involves both the S-resolvents:

Theorem 3.2.11 (The S-resolvent equation). *Let $T \in \mathcal{B}(\mathcal{X})$ and let $s, q \in \rho_S(T)$ with $q \notin [s]$. Then the equation*

$$S_R^{-1}(s, T)S_L^{-1}(p, T) = \left[\left(S_R^{-1}(s, T) - S_L^{-1}(q, T)\right) q \right.$$
$$\left. -\bar{s}\left(S_R^{-1}(s, T) - S_L^{-1}(q, T)\right)\right](q^2 - 2\mathrm{Re}(s)q + |s|^2)^{-1} \quad (3.13)$$

holds true.

Remark 3.2.12. We point out another main difference with respect to the classical operator theory: if A is a complex linear operator on a complex Banach Y space the resolvent set and the spectrum are associated with the invertibility of the operator $\lambda I - A$ and the operator $(\lambda I - A)^{-1} : \rho(A) \to B(Y)$ is a holomorphic function with values in the set of all bounded linear operators $B(Y)$. In the quaternionic setting the S-resolvent set and the S-spectrum are associated with the invertibility of $\mathcal{Q}_s(T) := T^2 - 2\mathrm{Re}(s)T + |s|^2 I$. However, the pseudo-resolvent operator

$$(T^2 - 2\mathrm{Re}(s)T + |s|^2 I)^{-1} : \rho_S(T) \to \mathcal{B}(\mathcal{X})$$

is not slice hyperholomorphic. The slice hyperholomorphicity is associated with the S-resolvent operators. In fact, the left S-resolvent $S_L^{-1}(s, T)$ is a $\mathcal{B}(\mathcal{X})$-valued right slice hyperholomorphic function of the variable s on $\rho_S(T)$, while the right S-resolvent $S_R^{-1}(s, T)$ is a $\mathcal{B}(\mathcal{X})$-valued left slice hyperholomorphic function of the variable s on $\rho_S(T)$.

The fact that the S-resolvent operators are left or right slice hyperholomorphic allows to define the S-functional calculus. To this end, we need some more notations:

Definition 3.2.13. *Let $T \in \mathcal{B}(\mathcal{X})$. We denote by $\mathcal{SH}_L(\sigma_S(T))$, $\mathcal{SH}_R(\sigma_S(T))$ and $\mathcal{N}(\sigma_S(T))$ the set of all left, right and intrinsic slice hyperholomorphic functions f with $\sigma_S(T) \subset \mathcal{D}(f)$, respectively.*

Definition 3.2.14 *(The S-functional calculus). Let $T \in \mathcal{B}(\mathcal{X})$. For any function $f \in \mathcal{SH}_L(\sigma_S(T))$, we define*

$$f(T) := \frac{1}{2\pi} \int_{\partial(U \cap \mathbb{C}_j)} S_L^{-1}(s, T) \, ds_j \, f(s), \quad (3.14)$$

where j is an arbitrary imaginary unit and U is an arbitrary slice Cauchy domain containing $\sigma_S(T)$. For any $f \in \mathcal{SH}_R(\sigma_S(T))$, we define

$$f(T) := \frac{1}{2\pi} \int_{\partial(U \cap \mathbb{C}_j)} f(s) \, ds_j \, S_R^{-1}(s, T), \quad (3.15)$$

where j is an arbitrary imaginary unit and U is an arbitrary slice Cauchy domain.

The S-functional calculus is well defined because the above integrals do not depend on U and $j \in \mathbb{S}$.

Theorem 3.2.15. *Let $T \in \mathcal{B}(\mathcal{X})$. For any $f \in \mathcal{SH}_L(\sigma_S(T))$, the integral in (3.14) that defines the operator $f(T)$ is independent of the choice of the slice Cauchy domain U containing $\sigma_S(T)$ and of the imaginary unit $j \in \mathbb{S}$. Similarly, for any $f \in \mathcal{SH}_R(\sigma_S(T))$, the integral in (3.15) that defines the operator $f(T)$ is also independent of the choice of U and $j \in \mathbb{S}$.*

Remark 3.2.16. Thanks to the functional calculus we can define functions of an operator T. In particular, we can define $(I - pT)^{-\star_r}$ using the function $(1 - pq)^{-\star_r}$ (where the $\star_r$ is computed with respect to the variable p). Note that for $p \neq 0$ we have

$$(I - pT)^{-\star_r} = p^{-1}S_R(p, T),$$

moreover

$$(I - pT)^{-\star_r} = \sum_{n \geq 0} p^n T^n \quad \text{for } |p| \|T\| < 1.$$

For the sake of simplicity, in the sequel we will write $(I - pT)^{-\star}$. This function is left slice hyperholomorphic in p. It is also interesting to note that $S_L(p, T) = (pI - T)^{-\star_l}$ and $S_R(p, T) = (pI - T)^{-\star_r}$, where both the $\star$-inverses are computed with respect to the variable p.

Some of the results that we mention in this section are stated for two-sided quaternionic Banach spaces, even though later we will mainly work with Hilbert spaces. Moreover, sometimes the Banach spaces under consideration are not two-sided. In the following proposition we recall an extension result, see [17, Proposition 3.24], which is valid in a more general setting and will be useful in the sequel:

Proposition 3.2.17. *Let A be a bounded linear operator from a right-sided quaternionic Banach $\mathcal{P}$ space into itself, and let G be a bounded linear operator from $\mathcal{P}$ into $\mathcal{Q}$, where $\mathcal{Q}$ is a two-sided quaternionic Banach space. The slice hyperholomorphic extension of $G(I - xA)^{-1}$, $1/x \in \sigma_S(A) \cap \mathbb{R}$, is*

$$(G - \overline{p}GA)(I - 2\text{Re}(p)\,A + |p|^2 A^2)^{-1}.$$

Remark 3.2.18. We also note that the Identity Principle, see [23], implies that two slice hyperholomorphic functions defined on an s-domain and with values in a two-sided quaternionic Banach space $\mathcal{X}$ coincide if their restrictions to the real axis coincide. More in general, any real analytic function $f : [a, b] \subseteq \mathbb{R} \to \mathcal{X}$ can be extended to a function $\text{ext}(f)$ slice hyperholomorphic on an axially symmetric s-domain U containing $[a, b]$. The fact that the extension exists is assured by the fact that for any $x_0 \in [a, b]$ the function f can be written as $f(x) = \sum_{n \geq 0} x^n A_n$, $A_n \in \mathcal{X}$, and x such that $|x - x_0| < \varepsilon_{x_0}$ and thus $(\text{ext} f)(p) = \sum_{n \geq 0} p^n A_n$ for $|p - x_0| < \varepsilon_{x_0}$. Thus the claim holds setting $B(x_0, \varepsilon_{x_0}) = \{p \in \mathbb{H} : |p - x_0| < \varepsilon_{x_0}\}$ and $U = \cup_{x_0 \in [a,b]} B(x_0, \varepsilon_{x_0})$.

Remark 3.2.19. The function

$$k(p,q) = (\bar{p} + \bar{q})(|p|^2 + 2\mathrm{Re}(p)\bar{q} + \bar{q}^2)^{-1} \tag{3.16}$$

is slice hyperholomorphic in p and $\bar{q}$ on the left and on the right, respectively, in its domain of definition, i.e., for $p \notin [-\bar{q}]$ (or, equivalently, $q \notin [-\bar{p}]$). It is positive definite in the open half-space $\mathbb{H}_+$. Its associated reproducing kernel Hilbert space is the Hardy space $\mathbf{H}_2(\mathbb{H}_+)$ of the right half-space of quaternions with positive real part.

The function $k(p,q) = \sum_{a=0}^{\infty} p^a \bar{q}^a$ is positive definite in the quaternionic unit open ball $\mathbb{B}_1$, with associated reproducing kernel Hilbert space $\mathbf{H}_2(\mathbb{B}_1)$. The corresponding space $\mathbf{H}_2(\mathbb{B}_1, \mathcal{H})$ of $\mathcal{H}$-valued functions (where $\mathcal{H}$ is a quaternionic Hilbert space) plays a key role in interpolation theory and model theory; see Proposition 5.1.6 for the latter.

3.3   The Map ω and Applications

Let f be a slice hyperholomorphic function. For $p = z \in \mathbb{C}_i$, and selecting an imaginary unit $j \in \mathbb{S}$ such that j is orthogonal to i, we can write the restriction of f to $\mathbb{C}_i$ in the form

$$f(z) = F(z) + G(z)j, \tag{3.17}$$

where F and G are $\mathbb{C}_i$-valued and also analytic from $\mathbb{C}_i$ into itself, since f is slice hyperholomorphic. We define (see [10, p. 400])

$$\omega_i(f)(z) = \begin{pmatrix} F(z) & G(z) \\ -\overline{G(\bar{z})} & \overline{F(\bar{z})} \end{pmatrix}. \tag{3.18}$$

We note that, strictly speaking, the map ω_i depends also on the choice of $j \in \mathbb{S}$. In the sequel, as we did for the map χ, we will write ω instead of ω_i.
We recall that

$$\omega(f \star g) = \omega(f)\omega(g) \tag{3.19}$$

and that ω coincides with the map χ when f is constant, so that, in particular:

$$\omega(fc) = \omega(f)\chi(c), \qquad c \in \mathbb{H}.$$

Let now $K(p,q)$ be $\mathbb{H}^{n \times n}$-valued, positive definite in the axially symmetric domain U. Assume moreover that f is left slice hyperholomorphic in p and right slice hyperholomorphic in $\bar{q}$, and let $\mathcal{H}(K)$ denote the associated reproducing kernel Hilbert space. It is separable in view of the slice hyperholomorphicity. Let $e_1, e_2, \ldots$ be an orthonormal basis of $\mathcal{H}(K)$. One can write the reproducing kernel as

$$K(p, q) = \sum_{u=1}^{\infty} e_u(p)e_u(q)^*,$$

and a function f belongs to $\mathcal{H}(K)$ if and only if it can be written as

$$f(p) = \sum_{u=1}^{\infty} e_u(p)c_u \qquad (3.20)$$

where $c_1, c_2, \ldots \in \mathbb{H}^n$ and are such that

$$\sum_{u=0}^{\infty} c_u^* c_u < \infty.$$

Applying the map ω to (3.20) we obtain

$$\omega(f) = \sum_{u=1}^{\infty} \omega(e_u)\chi(c_u). \qquad (3.21)$$

Theorem 3.3.1. *Let* $K(p, q) = \sum_{u=1}^{\infty} e_u(p)e_u(q)^*$.
(1) The function

$$\sum_{u=0}^{\infty} (\omega(e_u)(z)) \, (\omega(e_u)(w))^* \qquad (3.22)$$

is positive definite on $\mathbb{C}_i$.
(2) The associated reproducing kernel Hilbert space of $\mathbb{C}^{2n}$*-valued functions is the set of functions of the form*

$$F(z) = \sum_{u=0}^{\infty} \omega(e_u)d_u, \qquad (3.23)$$

where $d_0, d_1 \ldots \in \mathbb{C}^{2n}$, *such that*

$$\sum_{u=0}^{\infty} d_u^* d_u < \infty, \qquad (3.24)$$

with norm which is the infimum of (3.24) *over all representations.*
(3) The associated reproducing kernel Hilbert $\chi(\mathbb{C}^{n \times n})$*-module of* $\mathbb{C}^{2n \times 2n}$ *functions is the set of all functions of the form*

$$F(z) = \sum_{u=0}^{\infty} \omega(e_u)D_u, \qquad (3.25)$$

where $D_0, D_1 \ldots \in \chi(\mathbb{C}^{n \times n})$, such that

$$\text{Tr}\left(\sum_{u=0}^{\infty} D_u^* D_u\right) < \infty, \tag{3.26}$$

and associated $\mathbb{C}^{2n \times 2n}$-valued form

$$[F, G] = \sum_{u=0}^{\infty} G_u^* D_u, \quad \text{where} \quad G(z) = \sum_{u=0}^{\infty} \omega(e_u) G_u. \tag{3.27}$$

Proof. The proof follows with standard arguments so we do not repeat it.

Remark 3.3.2. A similar theorem could be stated with the map χ instead of the map ω_i. The functions will not be analytic then.

3.4   Slice Hyperholomorphic Weights: Half-Space and Unit Ball Cases

As recalled in the introduction (see Remarks 1.1.2), it is of interest to find $\mathcal{H}(A, B)$ spaces isometrically included in a $\mathbf{L}_2(d\mu)$ space, where $d\mu$ is a positive measure on the unit circle or on the real line. We consider here this question in the quaternionic setting, and other two questions pop up: what is the backward-shift operator in this context and what are the measures to be considered. Let us begin wih the backward-shift operators. For f slice hyperholomorphic in U and $\mathbb{H}^n$-valued and for $x, a \in U \cap \mathbb{R}$, we define as custiomary

$$(R_a f)(x) = \frac{f(x) - f(a)}{x - a}.$$

This function has slice hyperholomorphic extension to $U \setminus \{a\}$ equal to

$$(R_a f)(p) = (p - a)^{-\star} \star (f(p) - f(a)) = (p - a)^{-1}(f(p) - f(a)), \tag{3.28}$$

for $a \in U \cap \mathbb{R}$. Since

$$(R_a)f(x) - (R_b)f(x) = (a - b)(R_a R_b)f(x)$$

(the proof is as in the complex-valued case since x, a and b are chosen real) we have

$$(R_a f)(p) - (R_b f)(p) = (a - b)(R_a R_b f)(p), \quad p \in U.$$

We now turn to the problem of the measures. These will be now on $i\mathbb{R}$, for some arbitrary but fixed $i \in \mathbb{S}$, or on the unit circle. Let dn be a $\mathbb{H}^{n \times n}$-valued positive measure and let $\mathcal{M}$ be a space of quaternionic slice hyperholomorphic functions isometrically included $\mathbf{L}_2(dn)$: for $f \in \mathcal{M}$ we have

$$\int_{\mathbb{R}} f(it)^* dn(t) f(it) < \infty. \tag{3.29}$$

We could apply the map χ, so that we get

$$\mathrm{Tr} \int_{\mathbb{R}} \chi(f(it))^* \chi(dn(t)) \chi(f(it)) < \infty, \tag{3.30}$$

where we consider the boundary value of the function

$$\chi(f(z)) = \begin{pmatrix} F(z) & G(z) \\ -\overline{G(z)} & \overline{F(z)} \end{pmatrix},$$

where we wrote $f(z) = F(z) + G(z)j$ with F, G holomorphic, see (3.17). However, the problem is that the function $z \mapsto \chi(f(z))$ is clearly not analytic in $\mathbb{C}_i$.
Instead, one could consider the map ω and note that for $z, a \in \mathbb{C}_i$ and $f = F + Gj$ we have

$$R_a f = R_a F + R_a G j$$

and so, for *real a*,

$$\omega(R_a f) = R_a(\omega(f)).$$

But there seems to be no direct connection between a natural norm for $\omega(f)$ and (3.29), so also this idea seems to be not useful.
We thus proceed along a different line and we present a key result which allows to make the connection, as in the complex setting, between the quaternionic counterparts of the $\mathcal{H}(A, B)$ spaces and $\mathcal{H}(\Theta)$ spaces. To this end, we fix a pair (i, j) of orthogonal elements in $\mathbb{S}$. Let W_+ be a $\mathbb{H}^{n \times n}$-valued function slice hyperholomorphic in an axially symmetric U which contains $i\mathbb{R}$, and such that $\det \omega(W_+) \not\equiv 0$. In view of (3.19), this condition implies that

$$W_+ \star f \equiv 0 \quad \Longleftrightarrow \quad f = 0 \tag{3.31}$$

where f is slice hyperholomorphic in U. Thus we can define an inner product on the space of functions slice hyperholomorphic in U which are such that

$$\int_{\mathbb{R}} \|((W_+ \star f)(p))_{|p=it}\|^2 dt < \infty, \tag{3.32}$$

by setting

$$\langle f, g \rangle = \int_{\mathbb{R}} ((W_+ \star g)(p))^*_{|p=it} ((W_+ \star f)(p))_{|p=it} dt. \tag{3.33}$$

We note that in the above formulas, we first compute the $\star$-product between two functions and then we restrict to $i\mathbb{R}$.

So we can now give the following:

Definition 3.4.1. *We denote by* $\mathbf{L}_2(W_+^* W_+, dt)$ *the closure of the space of slice hyperholomorphic functions satisfying* (3.32).

As the complex case already illustrates with $W_+ \equiv 1$, this set contains functions which are not, in general, slice hyperholomorphic.

Remark 3.4.2. We note that, in the complex setting case, analytic weights are of importance, and are related for example to spectral factorizations (see, e.g., [92]). The notion of spectral factorization also intervenes in Subsection 10.2.

We note that for real a and b (note also that we often omit to write the argument of the functions to ease the notation):

$$W_+ \star R_a f = \frac{1}{p-a}(W_+ \star (f - f(a))) \quad \text{and} \quad W_+ \star R_b g = \frac{1}{p-b}(W_+ \star (g - g(b)))$$

and, for $p = it$,

$$\frac{1}{p-a} + \frac{1}{\overline{p}-b} = -\frac{a+b}{(p-a)(\overline{p}-b)}.$$

Hence we can write

$$(W_+ \star g)^*(W_+ \star R_a f) + (W_+ \star R_b g)^*(W_+ \star f) + (a+b)(W_+ \star R_b g)^*(W_+ \star R_a f)$$
$$= (W_+ \star g)^* \frac{1}{p-a}(W_+ \star (f - f(a)) + (W_+ \star (g - g(b)))^* \frac{1}{\overline{p}-b} -$$
$$- (W_+ \star (g - g(b)))^* \left(\frac{1}{\overline{p}-b} + \frac{1}{p-a} \right) (W_+ \star (f - f(a)))$$
$$= (W_+ g(b))^*(W_+ \star R_a f) + (W_+ \star R_b g)^*(W_+ f(a)).$$

So

$$\langle R_a f, g \rangle + \langle f, R_b g \rangle + (a+b)\langle R_a f, R_b g \rangle = G(b)^* J_1 F(a) \tag{3.34}$$

with

$$J_1 = \begin{pmatrix} 0 & I_n \\ I_n & 0 \end{pmatrix},$$

and

$$F(a) = \begin{pmatrix} f(a) \\ f_-(a) \end{pmatrix}, \tag{3.35}$$

the function $f_-(a)$ being defined by

$$f_-(a) = \langle R_a f, 1 \rangle \tag{3.36}$$

in the scalar case, and by

$$c^* f_-(a) = \langle R_a f, c \rangle$$

in the matrix-valued case.

We define a new space consisting of pairs $F = \begin{pmatrix} f \\ f_- \end{pmatrix}$, $G = \begin{pmatrix} g \\ g_- \end{pmatrix}$ equipped with inner product

$$\langle F, G \rangle = \langle f, g \rangle. \tag{3.37}$$

Formula (3.34) and (3.37) give

$$\langle R_a F, G \rangle + \langle F, R_b G \rangle + (a+b)\langle R_a F, R_b G \rangle = G(b)^* J_1 F(a). \tag{3.38}$$

One recognizes f_- as the function introduced in [25] and (3.38) as the structural identity of de Branges (see [45, 48]) characterizing a certain family of reproducing kernel Hilbert spaces.

Theorem 3.4.3. *Let $\mathcal{M} \subset \mathbf{L}_2(W_+^* W_+, dt)$ be a Hilbert space of functions slice hyperholomorphic in the open axially symmetric domain U, and assume $U \cap \mathbb{R} \neq \emptyset$. Assume moreover that $R_a \mathcal{M} \subset \mathcal{M}$ for $a \in U \cap \mathbb{R}$. Then there exists a J_1-inner function Θ such that the reproducing kernel of $\mathcal{M}$ is equal to*

$$-\frac{\Theta_{11}(z)\Theta_{12}(w)^* + \Theta_{12}(z)\Theta_{11}(w)^*}{z + \overline{w}}.$$

Proof. We proceed in a number of steps.

STEP 1: *Let $a, b \in \mathbb{R}$. We first check that*

$$\frac{(p-a)^{-1}(f(p) - f(a)) - (p-b)^{-1}(f(p) - f(b))}{a - b} =$$
$$= (p-a)^{-1}\left((p-b)^{-1}(f(p) - f(b)) - (a-b)^{-1}(f(a) - f(b))\right)$$

It suffices to compare the coefficients of $f(p)$, $f(a)$ and $f(b)$ on both sides. For $f(p)$ we have

$$\frac{1}{a-b}\left((p-a)^{-1} - (p-b)^{-1}\right)$$

for the left side, and this is equal to $(p-a)^{-1}(p-b)^{-1}$, which is the coefficient of $f(p)$ on the right side. The coefficients of $f(a)$ and $f(b)$ are treated in the same way.

STEP 2: *We note that*

$$(R_b f_-)(a) = ((R_b f)_-)(a).\tag{3.39}$$

We need to check that

$$\int_{\mathbb{R}} W_+(it)^* \left[W_+(p) \star (p-a)^{-1}(R_b f(p) - R_b f(a)) \right]_{p=it} dt = \frac{f_-(a) - f_-(b)}{a-b}.\tag{3.40}$$

Since

$$\frac{f_-(a) - f_-(b)}{a-b} = \frac{1}{a-b} \left(\int_{\mathbb{R}} W_+(it))^* \left[W_+(p) \star (p-a)^{-1}(f(p) - f(a)) \right]_{p=it} dt - \right.$$

$$\left. - \int_{\mathbb{R}} W_+(it))^* \left[W_+(p) \star (p-b)^{-1}((f(p) - f(b)) \right]_{p=it} dt \right)$$

we have (3.4), after removing the integrals and the multiplications by W_+ and W_+^*.

STEP 3: *The set $\mathcal{M}^\square$ of functions F of the form given in (3.35) endowed with the inner product (3.37) is a Hilbert space of slice hyperholomorphic functions which is R_b-invariant for real $b \in U$ and satisfies (3.38).*
This follows directly from Step 2. As in [25] we have

$$R_b F = R_b \begin{pmatrix} f \\ f_- \end{pmatrix} = \begin{pmatrix} R_b f \\ R_b(f_-) \end{pmatrix} = \begin{pmatrix} R_b f \\ (R_b f)_- \end{pmatrix}.$$

As in [25, Theorem 3.1 p. 604] the definition of inner product (3.37) allows us to show that the Hilbert space of functions F of the form (3.35) satisfy (3.38). The proof is then concluded by using the counterpart of [25, Theorem 3.1 p. 604], which is proved in [38].

Definition 3.4.4. *We will denote by $\mathcal{M}^\square$ the extension of the space $\mathcal{M}$ associated to the weight $W_+^* W_+$.*

Remark 3.4.5. More generally, one can consider inner products of the form

$$\langle f, g \rangle = \int_{\mathbb{R}} ((W_+ \star g)(p))^*_{|p=it}((W_+ \star f)(p))_{|p=it} d\mu(t)\tag{3.41}$$

where $d\mu$ is a scalar positive measure on the real line.

Finally, we discuss the quaternionic unit ball case. An analytic weight will now be a $\mathbb{H}^{n \times n}$-valued function, invertible and slice hyperholomorphic in a neighborhood of the closed quaternionic unit ball. We associate with the weight $W_+^* W_+$ the set of slice hyperholomorphic functions f such that

$$\int_0^{2\pi} \|((W_+ \star f)(p))_{|p=e^{it}}\|^2 dt < \infty.$$

We set

$$f_-(a) = \int_0^{2\pi} W_+(e^{it})^* \left((2pf(p) - pf(a) - af(a)) \star (p - a)^{-1} \star W_+(p) \right)_{|p=e^{it}} dt,$$

(compare with (3.36)). We note that on $(-1, 1)$ we have

$$f_-(x) = R_a(xf(x)) + xR_a f \qquad (3.42)$$

Proposition 3.4.6. *Let f, f_- be as above and $b \in \mathbb{R}$, then we have:*

$$(R_b f)_- = R_b f_-.$$

Proof. In view of (3.42) it is enough to prove that

$$R_b(R_a(xf(x)) + xR_a f) = R_a(R_b(xf(x)) + xR_b f),$$

that is, since R_a and R_b commute,

$$R_a(xR_b f) = R_b(xR_a f),$$

i.e.,

$$\frac{x\dfrac{f(x) - f(b)}{x - b} - a\dfrac{f(a) - f(b)}{a - b}}{x - a} = \frac{x\dfrac{f(x) - f(a)}{x - a} - b\dfrac{f(b) - f(a)}{b - a}}{x - b}.$$

It is immediate to check that the latter is an identity. $\qquad\square$

Chapter 4
Rational Functions

The notion of a rational slice hyperholomorphic function was introduced and studied in [23]. These functions play an important role in various aspects of quaternionic operator theory. It is always of interest to study the finite dimensional version of general results, and rational functions intervene in such cases (see, e.g., Corollaries 6.1.11 and 6.1.12 in Chapter 6).

4.1 Rational Slice Hyperholomorphic Functions

This chapter focuses on slice hyperholomorphic rational functions. Various equivalent definitions can be used to introduce them. The simplest seems to be the following:

Definition 4.1.1. *The $\mathbb{H}^{u \times v}$-valued function slice hyperholomorphic in an axially symmetric open set- $U \subseteq \mathbb{H}$ is rational if the function*

$$x \mapsto \chi(R(x))$$

is a rational function of $x \in U \cap \mathbb{R}$.

Following the realization theorem for complex-valued rational functions, one then has:

Theorem 4.1.2. *Let U and R be as above, and assume that $0 \in U$. Then, R is rational if and only if it can be written as*

$$R(p) = D + pC \star (I_m - pA)^{-\star} B \tag{4.1}$$

where $D = R(0)$ and $(A, B, C) \in \mathbb{H}^{m \times m} \times \mathbb{H}^{m \times v} \times \mathbb{H}^{u \times m}$.

© The Author(s), under exclusive license to Springer Nature Switzerland AG 2020
D. Alpay et al., *Quaternionic de Branges Spaces and Characteristic Operator Function*, SpringerBriefs in Mathematics,
https://doi.org/10.1007/978-3-030-38312-1_4

Definition 4.1.3. *Expression* (4.1) *is called a realization of R; it is called a minimal realization if m in* (4.1) *is minimal.*

When $(B, C) \in \mathbb{H}^{m \times v} \times \mathbb{H}^{u \times m}, u = v$ and D is invertible we have a formula expressing the $\star$-inverse of R:

$$R^{-\star}(p) = D^{-1} - pD^{-1}C \star (I_m - p(A - BD^{-1}C))^{-\star}BD^{-1}. \qquad (4.2)$$

Definition 4.1.4. *Let* $J \in \mathbb{R}^{u \times u}$ *be a signature matrix, i.e.,* $J = J^* = J^{-1}$, *and let* Θ *be a* $\mathbb{H}^{u \times u}$ *slice hyperholomorphic rational function. We say that* Θ *is* J-*unitary if*

$$\Theta(x)J\Theta(-x)^* = J \qquad (4.3)$$

at all real points where it is defined.
We say that a matrix D is J-*unitary if* $DJD^* = J$.

The following theorem is taken from [23, Theorem 9.3.1,p. 251] and is the counterpart of a result proved in [34] in the complex setting.

Theorem 4.1.5. *Let U be an axially symmetric open set in* $\mathbb{H}$ *with* $0 \in U$. *The function* Θ *is a slice hyperholomorphic rational function in U, with minimal realization*

$$\Theta(p) = D + pC \star (I - pA)^{-\star}B. \qquad (4.4)$$

if and only if the matrix D is D is J unitary, and there exists a uniquely determined Hermitian matrix H such that

$$C = DJB^*H, \qquad (4.5)$$
$$HA + A^*H = C^*JC. \qquad (4.6)$$

When $J = I_n$, rational matrix-valued unitary functions slice hyperholomorphic in the right half-space are finite Blaschke products. Let S be a $\mathbb{H}^{n \times n}$-valued unitary rational function, i.e., let S satisfy $S(x)S(-x)^* = I_n$ at those real points x where the expression makes sense. Then S can be written as

$$S(p) = B_1(p)^{-\star} \star B_2(p) \qquad (4.7)$$

where B_1 and B_2 are $\mathbb{H}^{n \times n}$-valued finite Blaschke products, see [19]. This is the quaternionic version of a special case of a general result of Krein and Langer (see [89] for the latter and [33] for its rational complex-valued version).
More generally we will need the following definition:

Definition 4.1.6. *The function B slice hypermeromorphic in the open unit ball* $\mathbb{B}_1$ *is called a Blaschke-Potapov product of the first kind (resp. second kind, resp. a singular factor) if* $\chi(B)(x)$, $x \in (-1, 1)$, *is the restriction to* $(-1, 1)$ *of a Blaschke-Potapov product of the first kind (resp. second kind, resp. singular factor) in* $\mathbb{B}_1$.

We now turn to the notion of Wiener-Hopf factorization.

Definition 4.1.7. *The $\mathbb{H}^{n \times n}$-valued rational function R slice hyperholomorphic in a neighborhood of infinity, with $R(\infty)$ invertible is said to admit a left Wiener-Hopf factorization if it can be written as $R = R_1 \star R_2$ where R_1 and R_2 are also $\mathbb{H}^{n \times n}$-valued rational functions slice hyperholomorphic and invertible in a neighborhood of infinity, and such that R_1 (resp. R_2) and its slice hyperholomorphic inverse have no poles in the half-space $\mathrm{Re}\, p \geq 0$ (resp. in $\mathrm{Re}\, p \leq 0$).*

Theorem 4.1.8. *Let R be a $\mathbb{H}^{n \times n}$-valued rational function such that $R(p) > 0$ for p in $\partial \mathbb{B}_1$, the boundary of the unit ball $\mathbb{B}_1$. Then R admits a Wiener-Hopf factorization of the form*

$$R(p) = R_+(p) \star R_+^c(p),$$

for a suitable function R.

The above result follows from [15] for the scalar case and from [108] for the matrix-valued case since a rational function invertible on $\partial \mathbb{B}_1$ belongs to the Wiener algebra. We will see an example of such factorization in Theorem 10.2.3.

Remark 4.1.9. In the case of Fueter variables, rational functions are defined in a similar way and appear first in the work [90]. They have been studied in [39] and [40].

4.2 Symmetries

The present section is in the complex setting, and will be used to prove Potapov's factorization theorem (Theorem 7.1.1 below) in the quaternionic setting.
The factorization of J-inner functions originates with the work of Potapov; see also [75], and [34] for the case of J-unitary rational functions. Following [7], the corresponding problem here is to consider minimal factorization of J-inner (and J-unitary) functions into factors which are J-inner (or J-unitary) and satisfy moreover the symmetry (4.8) below.
We will set

$$E = \begin{pmatrix} 0 & I_n \\ -I_n & 0 \end{pmatrix},$$

and we will consider a $\mathbb{C}^{2n \times 2n}$-valued rational function X such that

$$X(z) = E\overline{X(\overline{z})}E^{-1}, \quad z \in \mathbb{C}. \tag{4.8}$$

For such an X, we define the symmetry α by setting:

$$\alpha(X)(x) = E\overline{X(x)}E^{-1}, \quad x \in \mathbb{R}. \tag{4.9}$$

Lemma 4.2.1.
(1) *The symmetry α is multiplicative, i.e.,*

$$\alpha(XY) = \alpha(X)\alpha(Y),$$

for arbitrary $\mathbb{C}^{2n \times 2n}$-valued rational function X and Y, and satisfies

$$\alpha(X^*) = (\alpha(X))^*. \tag{4.10}$$

(2) *Let J be a signature matrix which commutes with E. If X is a product of Blasckhe-Potapov factors of the first kind (resp. second kind, third kind) so is $\alpha(X)$.*

Proof.
(1) Let X, Y be $\mathbb{C}^{2n \times 2n}$-valued rational functions. We have (where we remove the dependance on the variable to lighten the notation)

$$\begin{aligned}
\alpha(XY) &= E\overline{XY}E^{-1} \\
&= E\overline{X} \cdot \overline{Y}E^{-1} \\
&= E\overline{X}E^{-1}E\overline{Y}E^{-1} \\
&= \alpha(X)\alpha(Y).
\end{aligned}$$

Furthermore

$$\alpha(X^*) = E\overline{\overline{X^t}}E^{-1} = EX^tE^{-1}$$

while

$$(\alpha(X))^* = E^{-*}\overline{X}^*E^* = EX^tE^{-1}$$

since $\overline{X}^* = \overline{\overline{X^t}}$.
(2) This claim follows from

$$\begin{aligned}
\alpha\left(\frac{J - R(x)JR(y)^*}{1 - xy}\right) &= \frac{\alpha(J) - \alpha(R(x))\alpha(J)\alpha(R(y)^*)}{1 - xy} \\
&= \frac{\alpha(J) - \alpha(R(x))\alpha(J)(\alpha(R(y)))^*}{1 - xy} \\
&= \frac{J - \alpha(R(x))J(\alpha(R(y)))^*}{1 - xy},
\end{aligned}$$

since, by hypothesis, E commutes with J and so we deduce $\alpha(J) = EJE^{-1} = JEE^{-1} = J$. $\square$

We will be interested in slice hyperholomorphic rational functions R such that $X = \chi(R)$ satisfies the symmetry

$$X(x) = \alpha(X)(x), \quad x \in \mathbb{R}. \tag{4.11}$$

The symmetry can be written for a complex z, taking into account that analyticity must be preserved, as follows:

$$X(z) = E\overline{X(\overline{z})}E^{-1}, \quad z \in \mathbb{C}. \tag{4.12}$$

We now prove the following:

Theorem 4.2.2. *Let $X(z)$ be a $\mathbb{C}^{2n \times 2n}$-valued rational function, analytic at infinity, and with minimal realization*

$$X(z) = D + C(zI_u - A)^{-1}B.$$

Then, X satifies (4.12) if and only if there exists a uniquely defined matrix $S \in \mathbb{C}^{u \times u}$ such that

$$\overline{S}S = -I_u \tag{4.13}$$

and

$$\begin{pmatrix} A & B \\ C & D \end{pmatrix} = \begin{pmatrix} S & 0 \\ 0 & E \end{pmatrix} \begin{pmatrix} \overline{A} & \overline{B} \\ \overline{C} & \overline{D} \end{pmatrix} \begin{pmatrix} S^{-1} & 0 \\ 0 & E^{-1} \end{pmatrix}. \tag{4.14}$$

Proof. The function $E\overline{X(\overline{z})}E^{-1}$ is rational, analytic at infinity, and with minimal realization

$$E\overline{X(\overline{z})}E^{-1} = E\overline{D}E^{-1} + E\overline{C}(zI_u - \overline{A})^{-1}\overline{B}E^{-1}.$$

By uniqueness up to a similarity matrix of the minimal realization of a rational function analytic at infinity (see, e.g., [43]) there exists a uniquely defined matrix S such that (4.14) holds. Taking the conjugate of this equality and the fact that $E^{-1} = -E$ we see that it is also satisfied by $-\overline{S}^{-1}$, and hence the result. $\square$

We note that this result result is a special case of results in [6]. We also note that (4.14) can be rewritten in terms of A, B, C, D:

$$A = S\overline{A}S^{-1} \tag{4.15}$$
$$B = S\overline{B}E^{-1} \tag{4.16}$$
$$C = E\overline{C}S^{-1} \tag{4.17}$$
$$D = E\overline{D}E^{-1}. \tag{4.18}$$

Corollary 4.2.3. *An elementary factor satisfying the symmetry (4.8), normalized to be the identity at $z = 1$ and with singularities on the unit circle, is of the form*

$$X(z) = D \begin{pmatrix} \varphi(z)I & 0 \\ 0 & \overline{\varphi(z)}I \end{pmatrix} D^{-1} \tag{4.19}$$

where D satisfies (4.18) and

$$\varphi(z) = \frac{(z + e^{i\alpha})(1 - e^{-i\alpha})}{(z - e^{i\alpha})(1 + e^{i\alpha})}, \qquad \alpha \in \mathbb{R},$$

i.e., setting

$$D = \begin{pmatrix} D_1 & D_2 \\ -D_2 & D_1 \end{pmatrix}$$

we have

$$X(z) = \begin{pmatrix} D_1\overline{D_1}\varphi(z) + D_2\overline{D_2\varphi(z)} & -D_1 D_2(\varphi(z) + \overline{\varphi(z)}) \\ \overline{D_1 D_2}(\varphi(z) + \overline{\varphi(z)}) & D_1\overline{D_1\varphi(z)} + \overline{D_2}D_2\varphi(z) \end{pmatrix}. \qquad (4.20)$$

Proof. By hypothesis we can take $A = \mathrm{diag}(e^{i\alpha}, e^{-i\alpha})$, where $\alpha \in \mathbb{R}$. Condition (4.15) together with (4.13) implies that S is of the form

$$S = \begin{pmatrix} 0 & S_2 \\ -\overline{S_2}^{-1} & 0 \end{pmatrix},$$

and the statement follows.   □

Chapter 5
Operator Models

In [98], Rota studied models for linear operators in a Hilbert space: he proved that every linear operator T with spectral radius less than 1 is similar to the restriction of the adjoint of the unilateral shift S to a suitable invariant subspace. Thus, the unilateral shift S is what Rota called a "universal model" for such operator T. Intuitively, an operator S is a model for T when its restriction to one of its invariant subspaces is similar to T. In this short chapter we begin the study of operator models in the quaternionic framework.

5.1 Rota's Model in the Quaternionic Setting

In this section we discuss Rota's model for linear operators in quaternionic Hilbert spaces. Although the results and arguments are essentially the same as in the complex case, they rely on tools which have been developed only in recent years, and are specific to the quaternionic case. This allows to pinpoint important differences between the complex and quaternionic cases.
We begin by repeating some well-known notions in the quaternionic setting.

Definition 5.1.1. *A bounded linear operator T in a quaternionic Hilbert space $\mathcal{H}$ is said to be similar to an operator T' if there exists a bounded linear operator V with bounded inverse V^{-1} such that $T = V^{-1}T'V$.*

Definition 5.1.2. *A closed subspace $\mathcal{M}$ of $\mathcal{H}$ is said to be an invariant subspace of T as above if $Tx \in \mathcal{M}$ for every $x \in \mathcal{M}$.*

As it is well-known, invariant subspaces are important since they are related to the structure of an operator. We then give the following:

Definition 5.1.3. *Let T and S be a bounded linear operator in a quaternionic Hilbert space $\mathcal{H}$. We say that S is a model for T in $\mathcal{H}$ if there exists an invariant subspace*

© The Author(s), under exclusive license to Springer Nature Switzerland AG 2020   47
D. Alpay et al., *Quaternionic de Branges Spaces and Characteristic Operator Function*, SpringerBriefs in Mathematics,
https://doi.org/10.1007/978-3-030-38312-1_5

$\mathcal{M}$ of S and a bounded linear operator I_T in $\mathcal{H}$ whose range is the closed subspace $\mathcal{M}$ such that $T = I_T^{-1} S I_T$.

It is immediate that if S is a model for T and T' is similar to T then S is also a model for T'. We then recall that a bounded operator is said to be universal if for any bounded operator T', then some nonzero left multiple of T' is similar to a part of T, namely, there exists a nonzero $\lambda \in \mathbb{H}$, a closed subspace $\mathcal{M}$ of $\mathcal{H}$ invariant for T and a linear homeomorphisms ϕ of $\mathcal{H}$ onto $\mathcal{M}$ such that $\lambda T' = \phi^{-1}(T_{|\mathcal{M}})\phi$. Following Rota's paper [98], we can prove:

Theorem 5.1.4. *Let $\mathcal{H}$ be a right quaternionic Hilbert space. Then:*
(1) *Every bounded linear operator in $\mathcal{H}$ with S-spectrum in $\mathbb{B}_1$ is similar to a contraction of norm strictly less than 1.*
(2) *Every compact operator T with S-spectrum in $\overline{\mathbb{B}}_1$ and such that the norms of T^n are uniformly bounded is similar to a contraction.*
(3) *Every quasi-nilpotent operator is similar to a bounded operator of arbitrary small norm.*

Proof. We first prove the existence of a universal model for bounded operators in $\mathcal{H}$ whose S-spectrum is in $\mathbb{B}_1$. The proof follows as in [98], so we just repeat here the main lines. First of all, we consider the space $\ell_2(\mathbb{N}, \mathcal{H})$, which has the same dimension as $\mathcal{H}$ as shown by standard arguments. Since the S-spectrum is closed and the S-spectral radius $r_S(T)$ is the value of the spectrum of largest modulus, see Theorem 3.2.9 and so it is less than 1. In view of (3.12) the power series $\sum_{k=0}^{\infty} \|T^n x\|^2$ converges for every $x \in \mathcal{H}$. Furthermore, the space

$$\mathcal{M}(T) = \left\{ (x, Tx, T^2 x, \ldots), \ x \in \mathcal{H} \right\}$$

is closed in $\ell_2(\mathbb{N}, \mathcal{H})$. By the closed-graph theorem, see [21, 23], the map

$$\tau_T(x) = (x, Tx, T^2 x, \ldots)$$

is bounded invertible. Let F denote the forward shift in $\ell_2(\mathbb{N}, \mathcal{H})$, i.e.,

$$\mathsf{F}(x_1, x_2, x_3, \ldots) = (x_2, x_3, \ldots).$$

The operator F is a contraction, see [23]. We then have

$$T = \tau_T^{-1} \mathsf{F} \tau_T. \tag{5.1}$$

Since the Hilbert spaces $\mathcal{H}$ and $\ell_2(\mathbb{N}, \mathcal{H})$ have the same cardinality, one can find a unitary operator U from $\mathcal{H}$ onto $\ell_2(\mathbb{N}, \mathcal{H})$. Thus we can write

$$T = \tau_T^{-1} U^{-1} U \mathsf{F} U^{-1} U \tau_T. \tag{5.2}$$

The above argument shows that every linear operator T with S-spectrum inside the open unit ball is similar to a contraction. This contraction has possibly norm 1. With this result at hand we turn to the proof of (1). Replacing T by $(1 + \epsilon)T$ dilates the spectrum by a factor $(1 + \epsilon)$, and so it will stay inside $\mathbb{B}_1$ for $\epsilon > 0$ small enough. To prove (2) we reason as in the complex case, see [98].

So, we now turn to the last item. By hypothesis, for every $\epsilon > 0$ there exists $m \in \mathbb{N}$ such that

$$\|N^n\|^{1/n} \le \epsilon, \quad n \ge m.$$

Hence, $\left\| \frac{N^n}{\epsilon^n} \right\| \le 1$ for such n, and there exists $K > 0$ such that

$$\left\| \frac{N^n}{\epsilon^n} \right\| \le K, \quad n \in \mathbb{N}.$$

We can apply the result in item (2) to N/ϵ, and therefore N/ϵ is similar to a contraction S, which implies that N is similar to ϵS, which is of arbitrarily small norm. $\square$

Remark 5.1.5. Let F be the forward shift operator. The operator UFU^{-1} is a universal operator in the sense of Rota.

In the sequel we will be interested in the operator M_p of $\star$-multiplication by the quaternionic variable, and we note the following:

Proposition 5.1.6. *The operator F is unitarily equivalent to M_p from $\mathbf{H}_2(\mathbb{B}_1, \mathcal{H})$ into itself, and its adjoint is given by the backward-shift operator R_0.*

Proof. Both the assertions follows as in the complex case. $\square$

5.2 Operator Models

Consider a bounded linear operator T acting on a right quaternionic Pontryagin space $(\mathcal{P}, [\cdot, \cdot])$; we assume that its S-spectrum intersects the real line on an open set O. We note that for any $x \in \mathbb{R}$ the S-resolvent operator coincides with $(T - xI)^{-1}$. Assume that there is a $u \in \mathcal{P}$ such that the linear span of the vectors $(T^* - xI)^{-1}u$ is dense in $\mathcal{P}$ when x runs in O. The map $\mathscr{I}$ which to $f \in \mathcal{P}$ associates the function

$$F(x) = [f, (T^* - xI)^{-1}u] = [(T - xI)^{-1}f, u] \tag{5.3}$$

is one-to-one, and defines a Pontryagin space structure on the sets of such F. Let now $\lambda_0 \in O$. Using the resolvent identity we have:

$$
\begin{aligned}
[(T - x_0I)^{-1}f, (T^* - xI)^{-1}u] &= [(T - xI)^{-1}(T - x_0I)^{-1}f, u] \\
&= \frac{[(T - xI)^{-1}f, u] - [(T - x_0I)^{-1}f, u]}{x - x_0}.
\end{aligned}
$$

Thus

$$\mathscr{I}((T - x_0 I)^{-1} f) = R_{x_0}(\mathscr{I}(f)),$$

i.e., the resolvent operator $(T - x_0 I)^{-1}$ is unitarily equivalent to the backward-shift operator R_{x_0}. We note also that by Stone's theorem $R_{x_0} = (T - x_0 I)^{-1}$ when $\ker R_{x_0} = \{0\}$. This argument cannot be extended in a direct way when x, x_0 are not real.

Let $\mathcal{M}$ be a subspace of $\mathcal{P}$. If $f_0 \in \mathcal{M}$, $x, x_0 \in \mathbb{R}$ and $f_0(x_0) \neq 0$ we have

$$\frac{f(x) - f_0(x)\dfrac{f(x_0)}{f_0(x_0)}}{x - x_0} \in \mathcal{M}.$$

This can be written as

$$\frac{f(x) - f_0(x)\dfrac{f(x_0)}{f_0(x_0)}}{x - x_0} = f_0(x)\left(R_{x_0} f_0^{-1} f\right)(x).$$

We consider a (possibly) unbounded closed operator T for which there exists $u \in \mathcal{P}$ such that the elements

$$(xI - T)^{-*}u, \quad x \in \Omega \cap \mathbb{R},$$

are dense in $\mathcal{P}$ where Ω is an open set in $\mathbb{H}$ not intersecting the S-spectrum of T. When T is bounded, this condition is equivalent to ask that u is cyclic for T^*, that is the span of the vectors $u, T^*u, T^{*2}u, \ldots$ is dense in $\mathcal{P}$.

We denote by G the linear operator $Gf = [f, u]_{\mathcal{P}}$ and consider the function

$$L(x) = [(I - xT)^{-1}f, u] = G(I - xT)^{-1}f.$$

Assuming T bounded and using (3.2.17), the (unique) slice left-hyperholomorphic extension of L to a neighborhood of the origin is given by

$$L(p) = (G - \overline{p}GT)(I - 2(\operatorname{Re} p)T + |p|^2 T^2)^{-1} f. \tag{5.4}$$

We now define

$$F(p) = \frac{1}{p}L(1/p) = \frac{1}{p}\left(G - \frac{p}{|p|^2}GT\right)\left(I - 2\frac{\operatorname{Re} p}{|p|^2}T + \frac{1}{|p|^2}T^2\right)^{-1} f,$$

that is,

$$F(p) = \frac{1}{p}(|p|^2 G - pGT)(|p|^2 I - 2(\operatorname{Re} p)T + T^2)^{-1} f. \tag{5.5}$$

Note that $F(x) = G(xI - T)^{-1}f$ for $x \in \Omega \cap \mathbb{R}$. In view of the hypothesis on u, the function $F(p)$ is identically equal to 0 if and only if $f = 0$.

Assume we replace in (5.3) f by $S_L^{-1}(\alpha, T)f = (\alpha I - T)^{-1}f$. Then, the resolvent equation in Theorem 3.2.11 leads to

$$[S_R^{-1}(x, T)S_L^{-1}(\alpha, T)f, u]_{\mathcal{P}} =$$
$$= [(S_R^{-1}(x, T) - S_L^{-1}(\alpha, T))(x - \alpha)^{-1}f, u]_{\mathcal{P}}$$
$$= G(S_R^{-1}(x, T) - S_L^{-1}(\alpha, T))(x - \alpha)^{-1}f.$$

The left slice hyperholomorphic extension of this expression is

$$G \star (S_R^{-1}(p, T) - S_L^{-1}(T, \alpha)) \star (p - \alpha)^{-\star}f. \tag{5.6}$$

The proof of the next result follows easily from the previous arguments and will be omitted.

Theorem 5.2.1. *The space of functions F as in* (5.3) *endowed with the Hermitian form*

$$[F, F_1] = [f, f_1]_{\mathcal{P}} \tag{5.7}$$

is a reproducing kernel Pontryagin space with reproducing kernel $K(p, q)$ defined by

$$K(p, q) = G \star (pI - T)^{-\star}((qI - T)^{-\star})^* \star_r G^*, \tag{5.8}$$

where $\star$ and $\star_r$ are computed in p and q, respectively.

Chapter 6
Structure Theorems for $\mathcal{H}(A, B)$ Spaces

In the complex setting $\mathcal{H}(A, B)$ spaces, that is, reproducing kernel spaces with reproducing kernel of the form (1.4) (or their analogues with denominator replaced by $1 - z\overline{w}$) play an important role in the theory of operator models and related topics. In this section we focus on $\mathcal{H}(A, B)$ spaces in the quaternionic context.

6.1 $\mathcal{H}(A, B)$ Spaces

The following definition is the generalization to the quaternionic setting of the notion given in Section 1.1.

Definition 6.1.1. *A $\mathcal{H}(A, B)$ space is a reproducing kernel space of $\mathbb{H}$-valued (or more generally, $\mathbb{H}^n$-valued) functions, slice hyperholomorphic in an axially symmetric s-domain $\Omega \subseteq \mathbb{H}$, and with a reproducing kernel whose restriction to $\Omega \cap \mathbb{R}$ is of the form*

$$\frac{A(x)A(y)^* - B(x)B(y)^*}{x + y} \quad or \quad \frac{A(x)A(y)^* - B(x)B(y)^*}{1 - xy}, \qquad (6.1)$$

where A and B are $\mathbb{H}^{n \times n}$-valued functions slice hyperholomorphic in Ω.

One may wonder if the decomposition (6.1) is unique. To this end, we introduce the following:

Definition 6.1.2. *The pair (A, B) of $\mathbb{H}^{n \times n}$-valued functions slice hyperholomorphic in $\Omega \subseteq \mathbb{H}$ is full rank if the right linear span of the vectors*

$$\begin{pmatrix} A(x)^*c \\ B(x)^*c \end{pmatrix}, \quad x \in \Omega \cap \mathbb{R}, \quad c \in \mathbb{H}^n,$$

spans all of $\mathbb{H}^{2n}$.

D. Alpay et al., *Quaternionic de Branges Spaces and Characteristic Operator Function*, SpringerBriefs in Mathematics, https://doi.org/10.1007/978-3-030-38312-1_6

Proposition 6.1.3. *Assume the pair (A, B) in* (6.1) *to be full rank. Then, the pair (A, B) is unique up to a J_0-unitary constant factor, where*

$$J_0 = \begin{pmatrix} I & 0 \\ 0 & -I \end{pmatrix}.$$

Proof. We consider the kernel at the left hand side of (6.1). Let us assume that there exists a pair (A_1, B_1) such that

$$\frac{A(x)A(y)^* - B(x)B(y)^*}{x + y} = \frac{A_1(x)A_1(y)^* - B_1(x)B_1(y)^*}{x + y}, \quad x, y \in \Omega \cap \mathbb{R}.$$

Then,

$$A(x)A(y)^* - B(x)B(y)^* = A_1(x)A_1(y)^* - B_1(x)B_1(y)^*, \quad x, y \in \Omega \cap \mathbb{R}. \quad (6.2)$$

Let $x_1, \ldots, x_{2n} \in \Omega \cap \mathbb{R}$ and $c_1, \ldots, c_{2n} \in \mathbb{H}^n$ be such that the vectors

$$d_j = \begin{pmatrix} A(x_j)^* c_j \\ B(x_j)^* c_j \end{pmatrix}, \quad j = 1, \ldots, 2n,$$

span $\mathbb{H}^{2n}$, and let $D = \begin{pmatrix} d_1 & d_2 & \cdots & d_{2n} \end{pmatrix}$. We also define D_1 to be the corresponding matrix built from the pair (A_1, B_1). We have from (6.2)

$$D^* J_0 D = D_1^* J_0 D_1,$$

and the result follows since D is invertible. $\qquad\qquad\square$

Theorem 6.1.4. *Let Ω be an axially symmetric s-domain. Assuming in* (6.1) *that* $\det \chi(A) \not\equiv 0$, *we can rewrite*

$$\frac{A(x)A(y)^* - B(x)B(y)^*}{x + y} = A(x)\frac{I_n - S(x)S(y)^*}{x + y}A(y)^*. \quad (6.3)$$

with $S = A^{-1}B$. It follows that S is the restriction to $\Omega \cap \mathbb{R}$ of a Schur multiplier. If it is moreover inner, in the sense that the operator of multiplication by S is an isometry from $\mathbf{H}_2(\mathbb{H}_+)^n$ into itself, then

$$\mathcal{H}(A, B) = A \star \left(\mathbf{H}_2(\mathbb{H}_+)^n \ominus S \star \mathbf{H}_2(\mathbb{H}_+)^n \right), \quad (6.4)$$

and the following condition holds in $\mathcal{H}(A, B)$: for $p_0 \in \Omega \cap \mathbb{R}$, if $A(p_0)$ is invertible and $F \in \mathcal{H}(A, B)$ vanishes at $p_0 \in \Omega \cap \mathbb{R}$, then the function f_{p_0}

$$f_{p_0} : p \mapsto \frac{p + p_0}{p - p_0} \star f(p)$$

belongs to $\mathcal{H}(A, B)$ and has same norm as f:

$$\|f\| = \|f_{p_0}\|. \tag{6.5}$$

Proof. Assume S inner; then, $\mathbf{H}_2(\mathbb{H}_+)^n \ominus S \star \mathbf{H}_2(\mathbb{H}_+)^n$ is isometrically included in the Hardy space $\mathbf{H}_2(\mathbb{H}_+)^n$, and the restriction of its reproducing kernel to $(\Omega \cap \mathbb{R})^2$ is the function

$$\frac{I_n - S(x)S(y)^*}{x + y}.$$

Equality (6.4) follows then from [23]. Let now $F = A \star f \in \mathcal{H}(A, B)$ and such that $(A \star f)(p_0) = 0$ for some real $p_0 \in \Omega \cap \mathbb{R}$. Since p_0 is real we have

$$(A \star f)(p_0) = A(p_0)f(p_0)$$

and so $f(p_0) = 0$ since $A(p_0)$ is invertible. Thus

$$\frac{p + p_0}{p - p_0} \star (f(p) = (1 + 2p_0(p - p_0))^{-1} \star (f(p) - f(p_0)) \in \mathbf{H}_2(\mathbb{H}_+)^n \ominus S\mathbf{H}_2(\mathbb{H}_+)^n$$

since the space $\mathbf{H}_2(\mathbb{H}_+)^n \ominus S\mathbf{H}_2(\mathbb{H}_+)^n$ is R_{p_0}-invariant (see [23]). $\qquad\square$

Let us now consider a kernel as the one written on the right in (6.1).

Theorem 6.1.5. *Let Ω be an axially symmetric s-domain. Assuming in (6.1) that* $\det \chi(A) \not\equiv 0$ *we can rewrite*

$$\frac{A(x)A(y)^* - B(x)B(y)^*}{1 - xy} = A(x)\frac{I_n - S(x)S(y)^*}{1 - xy}A(y)^*. \tag{6.6}$$

with $S = A^{-1}B$. It follows that S is the restriction to $\Omega \cap \mathbb{R}$ of a Schur multiplier. If it is moreover inner, in the sense that the operator of multiplication by S is an isometry from $\mathbf{H}_2(\mathbb{B}_1)^n$ into itself. Then

$$\mathcal{H}(A, B) = A \star \left(\mathbf{H}_2(\mathbb{B}_1)^n \ominus S \star \mathbf{H}_2(\mathbb{B}_1)^n\right), \tag{6.7}$$

and the following condition holds in $\mathcal{H}(A, B)$: if $f \in \mathcal{H}(A, B)$ vanishes at $p_0 \in \Omega \cap \mathbb{R}$, then the function g_{q_0}

$$g_{q_0}: p \mapsto \frac{1 - pq_0}{p - q_0} \star f(p)$$

belongs to $\mathcal{H}(A, B)$ and has same norm as f:

$$\|f\| = \|g_{q_0}\|. \tag{6.8}$$

The proof is similar to the one of Theorem 6.1.4, and therefore omitted.

Theorem 6.1.6 (Half-space case). *The multiplication operator $M_p : f \mapsto pf$ is anti-hermitian from $\mathcal{H}(A, B)$ into itself.*

Proof. We have for f, g in the domain of M_p

$$\langle M_p f, g \rangle = \int_{\mathbb{R}} (g(it)^*(it) f(it) dt$$

$$= \int_{\mathbb{R}} (g(it)(-it))^* f(it) dt$$

$$= \langle f, -M_p g \rangle.$$

$\square$

Note that in the above statement, M_p need not be densely defined, let alone bounded, in fact we have:

Theorem 6.1.7. *In the notation of the previous theorems, the domain of M_p is given by the functions of the form $A \star f$, where $f \in \mathbf{H}_2(\mathbb{H}_+)^n \ominus S \star \mathbf{H}_2(\mathbb{H}_+)^n$ is such that $pf(p) \in \mathbf{H}_2(\mathbb{H}_+)^n$ and satisfies*

$$\langle f, Su \rangle = 0, \qquad u \in \mathbb{H}^n, \tag{6.9}$$

in the sense that

$$\langle (p + p_0) f, (p + p_0)^{-1} Su \rangle = 0, \tag{6.10}$$

where p_0 is arbitrary but fixed.

Proof. In view of the characterization (6.7), $pA \star f$ is in the domain of M_p (with $f \in \mathbf{H}_2(\mathbb{H}_+)^n \ominus S \star \mathbf{H}_2(\mathbb{H}_+)^n$ if and only if $pf \in \mathbf{H}_2(\mathbb{H}_+)$ and f is orthogonal to $S \star \mathbf{H}_2(\mathbb{H}_+)^n$. The latter condition is equivalent to

$$\langle pf(p), (p + p_0)^{-1} S(p)u \rangle = 0, \quad \forall p_0 > 0.$$

Rewriting

$$p(p + p_0)^{-1} = 1 - p_0(p + p_0)^{-1}$$

we have

$$\langle pf(p), (p + p_0)^{-1} S(p)u \rangle = \langle -f(p), (S(p) - p_0(p + p_0)^{-1} S(p))u \rangle = 0, \quad \forall p_0 > 0,$$

and so the result since the expression $\langle f(p), p_0(p + p_0)^{-1} S(p)u \rangle$ is well defined for all $p_0 > 0$. $\square$

When $\mathbb{H}_+$ is replaced by $\mathbb{B}_1$ we have:

Theorem 6.1.8 (Quaternionic unit ball case). *The operator* $M_p : f \mapsto pf$ *is a partial isometry from* $\mathcal{H}(A, B)$ *into itself.*

In the next two sections we study the converse statements: do the conditions imposed in the above two theorems force the form of the reproducing kernel? Before that we prove the following result in the case of $\mathbb{H}^n$-valued function is similar, but f_0 is now a matrix made of elements of $\mathcal{M}$ and a full rank hypothesis has to be added.

Proposition 6.1.9. *Let* $\mathcal{M}$ *be a space of* $\mathbb{H}^n$-*valued slice hyperholomorphic functions with the property that if* $f \in \mathcal{M}$ *and* $f(a) = 0$ *for* $a \in \Omega \cap \mathbb{R}$, *then the function*

$$p \mapsto (p - a)^{-1} f(p)$$

also belongs to $\mathcal{M}$. *Assume the following full rank hypothesis: there exist* $f_1, \ldots, f_n \in \mathcal{M}$ *such that* $\det \chi \left(f_1 \cdots f_n \right) \not\equiv 0$. *Let* $F = \left(f_1 \cdots f_n \right)$, *then the space* $F^{-1} \mathcal{M}$ *is* R_b-*invariant for* $b \in \Omega \cap \mathbb{R}$ *for which* $F(b)$ *is invertible.*

Proof. Indeed, let $g(x) = F(x)^{-1} f(x)$ (with extension $F(p)^{-\star} \star f(p)$). Then

$$R_b g(x) = \frac{F(x)^{-1} f(x) - F(b)^{-1} f(b)}{x - b}$$

$$= F(x)^{-1} \cdot \frac{f(x) - F(x)F(b)^{-1} f(b)}{x - b}.$$

To conclude we note that the function $x \mapsto f(x) - F(x)F(b)^{-1} f(b)$ belongs to $\mathcal{M}$, vanishes at $x = b$ and extends uniquely to a slice hyperholomorphic function belonging to $\mathcal{M}$. $\qquad\square$

Corollary 6.1.10. *Assuming that* $\mathcal{M}$ *in the previous proposition is finite dimensional, of dimension N, and assume regularity at the origin. The space* $\mathcal{M}$ *is the linear span of the columns of a matrix of the form*

$$F(p) \star C \star (I - pA)^{-\star}$$

where the pair $(C, A) \in \mathbb{H}^{n \times N} \times \mathbb{H}^{N \times N}$ *is observable, i.e.,*

$$\cap_{u=0}^{\infty} \ker C A^u = \{0\}. \qquad (6.11)$$

Proof. This follows from the structure of finite dimensional R_0-invariant spaces; see [23]. $\qquad\square$

In the next two corollaries we endow the space $\mathcal{M}$ with a (possibly indefinite) inner product defined by an Hermitian invertible matrix $P \in \mathbb{H}^{N \times N}$.

Corollary 6.1.11. *Assume that* P *is an invertible Hermitian matrix which satisfies*

$$P - A^* P A = C^* J C, \qquad (6.12)$$

where the matrix $I - A$ is assumed invertible. Then the reproducing kernel of the space $\mathcal{M}$ is of the form

$$\frac{J - \Theta(x)J\Theta(y)^*}{1 - xy}.$$

Proof. As in [23] (see, e.g., [5, Exercise 7.7.17, p. 368] for the case of complex functions) we define

$$\Theta(x) = I - (1 - x)C(I - xA)^{-1}P^{-1}(I - A)^{-*}C^*J.$$

Then,

$$C(I - xA)^{-1}P^{-1}(I - yA)^{-*}C^* = \frac{J - \Theta(x)J\Theta(y)^*}{1 - xy}.$$

□

Corollary 6.1.12. *Assume that P is an invertible Hermitian matrix which satisfies*

$$AP + PA^* + C^*JC = 0.$$

Then the reproducing kernel of $\mathcal{M}$ is of the form

$$F(x)\frac{J - \Theta(x)J\Theta(y)^*}{x + y}F(y)^*.$$

Proof. One now defines

$$\Theta(x) = I - C(xI - A)^{-1}P^{-1}C^*J.$$

Then,

$$C(xI - A)^{-1}P^{-1}(yI - A)^{-*}C^* = \frac{J - \Theta(x)J\Theta(y)^*}{x + y}$$

□

Remarks 6.1.13. If $J = I$ in the previous results we get $\mathcal{H}(A, B)$ spaces.

6.2 The Structure Theorem: Half-Space Case

Extending the theory of de Branges to the nonpositive case is an involved topic. We mention in particular the papers [85–87]. In this section and in the next one it will be easier to consider the quaternionic counterpart of (1.5) rather than the form (1.4) itself. The arguments still work for Krein spaces, and this is the setting we chose in this section and the following one. One should keep in mind that there is no one-to-one correspondence between reproducing kernel Krein spaces and difference of positive definite functions (see the discussion at the end of Section 2.3).

Lemma 6.2.1. *Let f be slice hyperhomolorphic in a neighborhood of the point $q_0 \in \mathbb{H}$ and assume that $f(q_0) = 0$. Then, $[q_0]$ is a removable singularity of the function*

$$g_{q_0}(p) = (p + \overline{q_0}) \star (p - q_0)^{-\star} \star f(p)$$

and g_{q_0} vanishes at the point $-\overline{q_0}$.

Proof. Indeed, we have

$$(p - q_0)^{-\star} = (p^2 - 2(\operatorname{Re} q_0)p + |q_0|^2)^{-1}(p - \overline{q_0}), \quad p \notin [q_0].$$

Furthermore, $f(p) = (p - q_0) \star h(p)$ where h is slice hyperholomophic in a neighborhood of q_0. Hence, for $p \notin [q_0]$ we have

$$\begin{aligned}
g_{q_0}(p) &= (p + \overline{q_0}) \star (p - q_0)^{-\star} \star f(p) \\
&= (p + \overline{q_0}) \star (p^2 - 2(\operatorname{Re} q_0)p + |q_0|^2)^{-1}(p - \overline{q_0}) \star (p - q_0) \star h(p) \\
&= (p + \overline{q_0}) \star h(p)
\end{aligned}$$

since $(p - \overline{q_0}) \star (p - q_0) = p^2 - 2(\operatorname{Re} q_0)p + |q_0|^2$. Thus g_{q_0} is slice hyperholomorphic in a neighborhood of q_0.
Furthermore, with $h(p) = \sum_{n=0}^{\infty} p^n h_n$ we have

$$(p + \overline{q_0}) \star h(p) = (p + \overline{q_0}) \star \left(\sum_{n=0}^{\infty} p^n h_n \right) = \sum_{n=0}^{\infty} p^n (p + \overline{q_0}) h_n,$$

and so $g_{q_0}(-\overline{q_0}) = 0$.   □

Conditions (6.15) and (6.24) below are restrictive. On the other hand, they are automatically satisfied in the positive case. In the statement below and in the sequel, $\widetilde{\mathbb{H}}$ denotes the set of purely imaginary quaternions.

Theorem 6.2.2. *Let $\mathcal{K}$ be a reproducing kernel Krein space of $\mathbb{H}$-valued functions slice hyperholomorphic in an axially symmetric s-domain Ω of the quaternions, and assume that the following condition holds: if $f \in \mathcal{K}$ vanishes at the point $q_0 \in \Omega \setminus \widetilde{\mathbb{H}}$, then*

$$f_{q_0} : p \mapsto f_{q_0}(p) = (p + \overline{q_0}) \star (p - q_0)^{-\star} \star f(p) \tag{6.13}$$

belongs to $\mathcal{K}$ and one has that

$$[f, f] = [f_{q_0}, f_{q_0}]. \tag{6.14}$$

Assume furthermore that there exists a real $p_0 \neq 0 \in \Omega$ such that

$$K(p_0, p_0)K(-p_0, -p_0) > 0. \tag{6.15}$$

Then, the reproducing kernel of $\mathcal{K}$ is of the form

$$K(p, q) = E_+(p) \star k(p, q) \star_r \overline{E_+(q)} - E_-(p) \star k(p, q) \star_r \overline{E_-(q)}, \tag{6.16}$$

where the functions E_+ and E_- are slice hyperholomorphic in Ω, and $k(p, q)$ is given by (see (3.16)):

$$k(p, q) = (\bar{p} + \bar{q})(|p|^2 + 2\mathrm{Re}(p)\bar{q} + \bar{q}^2)^{-1}.$$

Before proving the theorem we make some remarks.

Remarks 6.2.3.
(*a*) The conditions in the theorem are about zeros which are not in $\widetilde{\mathbb{H}}$, but elements in $\mathcal{K}$ may have zeros which are in $\widetilde{\mathbb{H}}$.
(*b*) The function (6.13) has a removable singularity at the sphere $[p_0]$. This has been explained in Lemma 6.2.1.
(*c*) In the positive case, condition (6.15) is automatically satisfied and, for entire complex functions, this theorem was first proved by L. de Branges; see [47, Theorem 23, p. 57]. We will give in Section 6.4 examples where (6.15) is in force.

Proof of Theorem 6.2.2 We divide the proof in a number of steps.

STEP 1: *Let $p_0 \in \Omega$ such that $K(p_0, p_0)$ is not zero. The function*

$$(p + \overline{p_0}) \star (p - p_0)^{-\star} \star \left(K(p, q) - \frac{K(p, p_0)K(p_0, q)}{K(p_0, p_0)} \right)$$

has a removable singularity at the sphere $[p_0]$, belongs to the space and vanishes at the point $-\overline{p_0}$.
This is a consequence of Lemma 6.2.1 and of the hypothesis of the theorem since the function

$$p \mapsto \left(K(p, q) - \frac{K(p, p_0)K(p_0, q)}{K(p_0, p_0)} \right) \tag{6.17}$$

clearly belongs to $\mathcal{K}$.
STEP 2: *Assume that there exists a nonzero $p_0 \in \Omega$ such that $K(p_0, p_0)$ and $K(-\overline{p_0}, -\overline{p_0})$ are both non zeros. Then for real s one obtains*

$$(p + \overline{p_0}) \star (p - p_0)^{-\star} \star \left(K(p, s) - \frac{K(p, p_0)K(p_0, s)}{K(p_0, p_0)} \right)$$

$$= \left(K(p, s) - \frac{K(p, -\overline{p_0})K(-\overline{p_0}, s)}{K(-\overline{p_0}, -\overline{p_0})} \right) (s - \overline{p_0})(s + p_0)^{-1}. \tag{6.18}$$

The argument is that of de Branges, adapted to the fact that we are in the quaternionic setting. The idea is to compute in two different ways the inner product

$$[f(p), (p + \overline{p_0}) \star (p - p_0)^{-*} \star \left(K(p, q) - \frac{K(p, p_0)K(p_0, q)}{K(p_0, p_0)} \right)]$$

where f vanishes at $p = -\overline{p_0}$. Note that the function

$$(p + \overline{p_0}) \star (p - p_0)^{-*} \star \left(K(p, q) - \frac{K(p, p_0)K(p_0, q)}{K(p_0, p_0)} \right)$$

vanishes also at the point $-\overline{p_0}$.

In the first way, we take $q = s$ to be real and make use of the isometry hypothesis (with the point $q_0 = -\overline{p_0}$) to write for real s

$$[f(p), (p + \overline{p_0}) \star (p - p_0)^{-*} \star \left(K(p, s) - \frac{K(p, p_0)K(p_0, s)}{K(p_0, p_0)} \right)] =$$

$$= [(p - p_0) \star (p + \overline{p_0})^{-*} \star f(p), K(p, s) - \frac{K(p, p_0)K(p_0, s)}{K(p_0, p_0)}]$$

$$= (s - p_0)(s + \overline{p_0})^{-1} f(s)$$

since the function $(p - p_0) \star (p + \overline{p_0})^{-*} \star f(p)$ vanishes at $p = p_0$ and hence

$$0 = [(p - p_0) \star (p + \overline{p_0})^{-*} \star f(p), \frac{K(p, p_0)K(p_0, s)}{K(p_0, p_0)}].$$

In the second way we rewrite $(s - p_0)(s + \overline{p_0})^{-1} f(s)$ as

$$(s - p_0)(s + \overline{p_0})^{-1} f(s) = [f(p), \left(K(p, s) - \frac{K(p, -\overline{p_0})K(-\overline{p_0}, s)}{K(-\overline{p_0}, -\overline{p_0})} \right) (s - \overline{p_0})(s + p_0)^{-1}].$$

(6.19)

Above we used the fact that $f(-\overline{p_0}) = 0$ and hence

$$0 = [f(p), \frac{K(p, -\overline{p_0})K(-\overline{p_0}, s)}{K(-\overline{p_0}, -\overline{p_0})}(s - \overline{p_0})(s + p_0)^{-1}].$$

We thus have for every function f vanishing at $-\overline{p_0}$

$$[f(p), (p + \overline{p_0}) \star (p - p_0)^{-*} \star \left(K(p, s) - \frac{K(p, p_0)K(p_0, s)}{K(p_0, p_0)} \right)] =$$

$$= [f(p), \left(K(p, s) - \frac{K(p, -\overline{p_0})K(-\overline{p_0}, s)}{K(-\overline{p_0}, -\overline{p_0})} \right) (s - \overline{p_0})(s + p_0)^{-1}].$$

STEP 3: *The reproducing kernel is of the form*

$$K(p,q) = \frac{1}{u}E(p) \star k(p,q) \star_r \overline{E(q)} - \frac{1}{v}F(p) \star k(p,q) \star_r \overline{F(q)} \qquad (6.20)$$

with k(p, q) as in (3.16), *and*

$$E(p) = (p + \overline{p_0}) \star K(p, p_0), \quad F(p) = (p - p_0) \star K(p, -\overline{p_0}),$$

and

$$u = \frac{2(\text{Re } p_0)}{K(p_0, p_0)} \quad \text{and} \quad v = \frac{2(\text{Re } p_0)}{K(-\overline{p_0}, -\overline{p_0})}. \qquad (6.21)$$

For real $p = t$, and multiplying equality (6.18) on the left by $(t - p_0)$ and on the right by $s + p_0$ we obtain:

$$(t + \overline{p_0})K(t,s)(s + p_0) - (t - p_0)K(t,s)(s - \overline{p_0}) =$$
$$= \frac{(t + \overline{p_0})K(t, p_0)K(p_0, s)(s + p_0)}{K(p_0, p_0)} - \frac{(t - p_0)K(t, -\overline{p_0})K(-\overline{p_0}, s)(s - \overline{p_0})}{K(-\overline{p_0}, -\overline{p_0})}.$$
$$(6.22)$$

Note that $K(t, s)$ commutes with $\overline{p_0}$ since it commutes with p_0. Thus we can rewrite (6.22) as

$$2(\text{Re } p_0)K(t,s)(t + s) = \frac{(t + \overline{p_0})K(t, p_0)K(p_0, s)(s + p_0)}{K(p_0, p_0)}$$

$$- \frac{(t - p_0)K(t, -\overline{p_0})K(-\overline{p_0}, s)(s - \overline{p_0})}{K(-\overline{p_0}, -\overline{p_0})}$$

and hence

$$K(t, s) = \frac{E(t)u\overline{E(s)} - F(t)v\overline{F(s)}}{t + s}, \quad t, s \in \Omega \cap \mathbb{R}.$$

The result follows by slice hyperholomorphic extension.

STEP 4: *The reproducing kernel is of the form* (6.16).

By hypothesis, $uv > 0$. It suffices to take

$$E_+(p) = \sqrt{u}E(p) \quad \text{and} \quad E_-(p) = \sqrt{v}F(p)$$

if $u > 0$, and

$$E_+(p) = \sqrt{-v}F(p) \quad \text{and} \quad E_-(p) = \sqrt{-u}E(p)$$

if $u < 0$. □

Remarks 6.2.4. Assume that $\mathcal{K}$ is a Pontryagin space. Then $uv > 0$ in the preceding proof. Indeed, assume $uv < 0$ If $u < 0$ and $v > 0$, the kernel $K(t,s)$ has an infinite number of negative squares on $\Omega \cap \mathbb{R}$, contradicting the hypothesis that $\mathcal{K}$ is a Pontryagin space.

If $u > 0$ and $v < 0$, the kernel $K(p,q)$ (equal to (6.20)) is positive definite, and in particular both $K(p_0, p_0)$ and $K(-\overline{p_0}, -\overline{p_0})$ are positive. This forces

$$uv = \frac{(\text{Re } p_0))^2}{K(p_0, p_0)K(-\overline{p_0}, -\overline{p_0})} > 0$$

which cannot be for $u > 0$ and $v < 0$.

Remarks 6.2.5. Assume that $K(p, -\overline{p_0}) \equiv 0$. Then the reproducing kernel is of the form (6.16) with $E_-(p) \equiv 0$.

As an example of a space where conditions (6.13) and (6.14) hold for real p_0, consider a slice hyperholomorphic function A. Then for any slice hyperholomorphic function f we have

$$A \star \frac{p + p_0}{p - p_0} \star f = \frac{p + p_0}{p - p_0} A \star f$$

and hence for $p = it$ we have

$$|(A \star \frac{p + p_0}{p - p_0} \star f)(p)| = |(A \star f)(p)|$$

and

$$\int_{\mathbb{R}} \left| (A \star \frac{p + p_0}{p - p_0} \star f)(p)|_{p=it} \right|^2 dt = \int_{\mathbb{R}} |(A \star f)(p)|_{p=it}|^2 dt.$$

6.3 The Unit Ball Case

Theorem 6.3.1. *Let $\mathcal{K}$ be a reproducing kernel Krein of $\mathbb{H}$-valued functions slice hyperhomorphic in an axially symmetric s-domain Ω of the quaternions, symmetric with respect to the unit sphere $\mathbb{H}_1$ of the quaternions (i.e., if $\beta \in \Omega \setminus \{0\}$ then $(\overline{\beta})^{-1} \in \Omega$) and assume that the following condition holds: If $f \in \mathcal{K}$ vanishes at the point $q_0 \in \Omega \setminus \mathbb{H}_1$ then the function f_{q_0} defined by*

$$p \mapsto (1 - p\overline{q_0}) \star (p - q_0)^{-\star} \star f(p) \tag{6.23}$$

belongs to $\mathcal{K}$ and

$$[f, f] = [f_{q_0}, f_{q_0}]. \tag{6.24}$$

Assume furthermore that there exists a point $p_0 \neq 0 \in \Omega$ such that (6.24) holds, as well as

$$K(p_0, p_0)K(\overline{p_0}^{-1}, \overline{p_0}^{-1}) > 0, \tag{6.25}$$

and

$$[F\frac{\overline{p_0}}{p_0}, F\frac{\overline{p_0}}{p_0}] = [F, F]. \tag{6.26}$$

Then, the reproducing kernel of $\mathcal{K}$ *is of the form*

$$K(p, q) = E_+(p) \star (1 - p\overline{q})^{-\star} \star_r \overline{E_+(q)} - E_-(p) \star (1 - p\overline{q})^{-\star} \star_r \overline{E_-(q)}, \tag{6.27}$$

where the functions E_+ *and* E_- *are slice hyperholomorphic in* Ω.

Remarks 6.3.2.
(*a*) The complex-valued version of the above theorem can be found in the Hilbert space case in [9, Theorem 6.1, p. 173]. In the case of spaces of polynomials it was proved earlier in [91, Theorem 1, p. 231].
(*b*) Equations (6.24) and (6.26) are satisfied in particular if p_0 can be chosen real. This will hold in particular in the Hilbert space case.
(*c*) As for the half-plane case, condition (6.27) will hold in the case of a Pontryagin space.

Proof of Theorem 6.3.1 We follow the proof as in [9], suitably modified to take into account the noncommutativity of the quaternions, as in Theorem 6.2.2 above.
Let p_0 be as in the statement of the theorem. The function (6.17) belongs to $\mathcal{K}$ and so does the function

$$(1 - p\overline{p_0}) \star (p - p_0)^{-\star} \star \left(K(p, q) - \frac{K(p, p_0)K(p_0, q)}{K(p_0, p_0)} \right). \tag{6.28}$$

This last function vanishes at the point $(\overline{p_0})^{-1}$. So, with

$$\Delta(p) = (1 - pp_0^{-1}) \star (p - (\overline{p_0})^{-1})^{-\star} \star (1 - p\overline{p_0}) \star (p - p_0)^{-\star} = \frac{\overline{p_0}}{p_0}$$

we can write for every $F \in \mathcal{K}$ vanishing at $(\overline{p_0})^{-1}$

$$[F(p), (1 - p\overline{p_0}) \star (p - p_0)^{-\star} \star \left(K(p, q) - \frac{K(p, p_0)K(p_0, q)}{K(p_0, p_0)} \right)] =$$

$$= [(p - (\overline{p_0})^{-1})^{-\star} \star (1 - pp_0^{-1}) \star F(p), \Delta(p) \star \left(K(p, q) - \frac{K(p, p_0)K(p_0, q)}{K(p_0, p_0)} \right)]$$

$$= [(p - p_0) \star (1 - p\overline{p_0})^{-\star} \star F(p)\frac{\overline{p_0}}{p_0}, \left(K(p, q) - \frac{K(p, p_0)K(p_0, q)}{K(p_0, p_0)} \right)\frac{\overline{p_0}}{p_0}]$$

$$= [(p - p_0) \star (1 - p\overline{p_0})^{-\star} \star F(p), \left(K(p, q) - \frac{K(p, p_0)K(p_0, q)}{K(p_0, p_0)} \right)]$$

$$= [(p - p_0) \star (1 - p\overline{p_0})^{-\star} \star F(p), K(p, q)] -$$

$$- [(p - p_0) \star (1 - p\overline{p_0})^{-\star} \star F(p), \frac{K(p, p_0)K(p_0, q)}{K(p_0, p_0)}]$$

where we have used (6.26). Taking $q = s$ real and taking into account that

$$(p - p_0) \star (1 - p\overline{p_0})^{-\star} \star F(p)$$

vanishes at p_0 we thus have

$$[F(p), (p - p_0) \star (1 - p\overline{p_0})^{-\star} \star \left(K(p, s) - \frac{K(p, p_0)K(p_0, s)}{K(p_0, p_0)} \right)] = (s - p_0)(1 - s\overline{p_0})^{-1} F(s).$$

On the other hand, since $F((\overline{p_0})^{-1}) = 0$, the reproducing kernel property gives

$$(s - p_0)(1 - s\overline{p_0})^{-1} F(s) = [F, \left(K(p, s) - \frac{K(p, (\overline{p_0})^{-1})K((\overline{p_0})^{-1}, s)}{K((\overline{p_0})^{-1}, (\overline{p_0})^{-1})} \right) (s - \overline{p_0})(1 - sp_0)^{-1}].$$

Hence we get

$$(1 - p\overline{p_0}) \star (p - p_0)^{-\star} \star \left(K(p, s) - \frac{K(p, p_0)K(p_0, s)}{K(p_0, p_0)} \right) =$$

$$= \left(K(p, s) - \frac{K(p, (\overline{p_0})^{-1})K((\overline{p_0})^{-1}, s)}{K((\overline{p_0})^{-1}, (\overline{p_0})^{-1})} \right) (s - \overline{p_0})(1 - sp_0)^{-1}.$$

Setting $p = t$ real, multiplying on the left by $(1 - p\overline{p_0})$ and on the right by $(1 - sp_0)$ we get:

$$(1 - t\overline{p_0}) \left(K(t, s) - \frac{K(t, p_0)K(p_0, s)}{K(p_0, p_0)} \right) (1 - sp_0) =$$

$$= (t - p_0) \left(K(t, s) - \frac{K(t, (\overline{p_0})^{-1})K((\overline{p_0})^{-1}, s)}{K((\overline{p_0})^{-1}, (\overline{p_0})^{-1})} \right) (s - \overline{p_0}).$$

Since (6.24) holds, we get the result as in the proof of Theorem 6.2.2.   □

6.4   The Conditions (6.15) and (6.25)

Assume that $K(p, q)$ is positive definite, and not identically vanishing. The Cauchy-Schwarz inequality will imply that (6.15) or (6.25) hold. A similar conclusion holds if we only know that $K(p_0, p_0) \neq 0$.

Proposition 6.4.1. *Assume $K(p, q) \neq 0$. There exist $p \in \Omega$ such that $K(p, p) \neq 0$.*

Proof. If the kernel is positive definite the claim follows trivially from the Cauchy-Schwarz inequality. In the general case, we adapt an argument from [30, Proof of Theorem 4.2, p. 50]. Let $x_0 \in \Omega \cap \mathbb{R}$, and let

$$K(p, q) = \sum_{n,m=0}^{\infty} (p - x_0)^n k_{n,m} (\overline{q} - x_0)^n$$

be the power series of $K(p, q)$ near the point (x_0, x_0), with quaternionic coefficients $k_{n,m}$. We write $k_{n,m} = a_{n,m} + b_{n,m} j$, where $a_{n,m}$ and $b_{n,m}$ are complex numbers in $\mathbb{C}_i$. To show that these numbers are equal to 0 we take various choices of p.

Case 1. $p = x_0 + re^{it}$ with $r > 0$ and $t \in \mathbb{R}$: Since

$$(a_{n,m} + b_{n,m} j)e^{-imt} = e^{-imt} a_{n,m} + e^{imt} b_{n,m} j,$$

the condition $K(p, p) \equiv 0$ can be rewritten as

$$\left(\sum_{n,m=0}^{\infty} r^{n+m} e^{i(n-m)t} a_{n,m} \right) + \left(\sum_{n,m=0}^{\infty} \rho^{n+m} e^{i(n+m)t} b_{n,m} \right) j \equiv 0,$$

and it follows that

$$\sum_{n,m=0}^{\infty} r^{n+m} e^{i(n-m)t} a_{n,m} \equiv 0,$$

The complex case argument gives then $a_{n,m} = 0$ for all $n, m \in \mathbb{N}_0$. Note that the vanishing of the factor of j will not lead to the similar conclusion for the $b_{n,m}$.

Case 2. $p = x_0 + re^{jt}$: In view of the first case, $K(p, p)$ can be rewritten as

$$\sum_{n,m=0}^{\infty} p^n b_{n,m} j \overline{p}^m = \left(\sum_{n,m=0}^{\infty} r^{n+m} (\operatorname{Re} b_{n,m}) e^{j(n-m)t} \right) j$$

$$+ \left(\sum_{n,m=0}^{\infty} r^{n+m} e^{jnt} i (\operatorname{Im} b_{n,m}) e^{-jmt} \right) j$$

$$= \left(\sum_{n,m=0}^{\infty} r^{n+m} (\operatorname{Re} b_{n,m}) e^{j(n-m)t} \right) j$$

$$+ \left(\sum_{n,m=0}^{\infty} r^{n+m} e^{j(n+m)t} (\operatorname{Im} b_{n,m}) \right) ij.$$

Thus $K(p, p) \equiv 0$ leads to

$$\left(\sum_{n,m=0}^{\infty} r^{n+m} (\operatorname{Re} b_{n,m}) e^{j(n-m)t} \right) \equiv 0,$$

and we have that $\operatorname{Re} b_{n,m} = 0$ for all $n, m \in \mathbb{N}_0$.

Case 3. $p = x_0 + re^{kt}$ (with $k = ij$): Then

$$\sum_{n,m=0}^{\infty} p^n ij \, (\operatorname{Im} b_{n,m}) \overline{p}^m = k \cdot \sum_{n,m=0}^{\infty} \rho^{n+m} e^{k(n-m)t} \, (\operatorname{Im} b_{n,m})$$

and so $\operatorname{Im} b_{n,m} \equiv 0$.

$\square$

6.5 A Theorem on the Zeros of a Polynomial

We now turn to the counterpart of Theorem 2.2.5.

Theorem 6.5.1. *Let $T \in \mathbb{H}^{n \times n}$ be an invertible Hermitian matrix, with $v \geq 0$ negative eigenvalues. Assume furthermore that*

$$\left(1 \; x \; \ldots \; x^n\right) T^{-1} \left(1 \; y \; \ldots \; y^n\right)^t = \frac{A(x)\overline{A(y)} - x\overline{y}B(z)\overline{B(w)}}{1 - xy} \tag{6.29}$$

where A and B are polynomials of degree n. Then, A has v zeros inside $\mathbb{B}_1$ and B has v zeros outside $\mathbb{B}_1$. They have no zeros on the boundary $\partial \mathbb{B}_1$.

We remark that (6.29) can be rewritten as

$$\left(1 \; p \; \ldots \; p^n\right) T^{-1} \left(1 \; q \; \ldots \; q^n\right)^* = \left(A(p)\overline{A(q)} - pB(p)\overline{B(q)\overline{q}}\right) \star (1 - p\overline{q})^{-\star}$$

$$= \sum_{u=0}^{\infty} p^u \left(A(p)\overline{A(q)} - pB(p)\overline{B(q)\overline{q}}\right) \overline{q}^u \tag{6.30}$$

by slice hyperholomorphic extension.

Proof of Theorem 6.5.1 We follow the steps of the proof of Theorem 2.2.5. The first step still holds since the spectral theorem holds for quaternionic Hermitian matrices. The second step works also in the same way. To consider the third step in the quaternionic setting, we first rewrite (6.30) as

$$A(p)\overline{A(q)} - pB(p)\overline{B(q)}\overline{q} = \left(1 \; p \; \ldots \; p^n\right) T^{-1} \left(1 \; q \; \ldots \; q^n\right)^*$$
$$- p \left(1 \; p \; \ldots \; p^n\right) T^{-1} \left(1 \; q \; \ldots \; q^n\right)^* \overline{q}. \tag{6.31}$$

Let q_0 be a zero of (say A) on $\partial \mathbb{B}_1$. Remark that

$$\left(1 \; q_0 \; \ldots \; q_0^n\right) T^{-1} \left(1 \; q_0 \; \ldots \; q_0^n\right)^* \in \mathbb{R} \tag{6.32}$$

and so setting $p = q = q_0$ in (6.31) leads to $B(q_0) = 0$, contradicting the previous step. The next two steps are also the same since the Krein-Langer factorization theorem holds in the slice hyperholomorphic setting. □

Remarks 6.5.2. The argument on the lack of zeros on the boundary $\partial\mathbb{B}_1$ will not hold in the matrix-valued case; then, (6.32) is not real but is a quaternionic Hermitian matrix which will not commute with q_0 in general.

Chapter 7
J-Contractive Functions

As explained in the introduction there are (at least) three important families of reproducing kernel Hilbert spaces introduced by de Branges and Rovnyak, and used in operator models and related topics. In the previous section we studied the quaternionic counterpart of $\mathcal{H}(A, B)$ spaces. In this section and in the next one we study $\mathcal{H}(\Theta)$ spaces in this setting.

7.1 *J*-Contractive Functions in the Quaternionic Unit Ball

Recalling the definition of hypermeromorphic functions, see Definition 3.2.4, we can prove our next result:

Theorem 7.1.1. *Let Θ be a $\mathbb{H}^{n \times n}$-valued function slice hypermeromorphic in $\mathbb{B}_1$ and J-contractive there, and slice hyperholomorphic in a neighborhood of $p = 1$. Then it can be written as*

$$\Theta(p) = \Theta_1(p) \star \Theta_2(p) \star \Theta_3(p), \tag{7.1}$$

where Θ_1 is a Blaschke-Potapov product with all its zeros inside the open unit ball, Θ_2 is a Blaschke-Potapov product with all its zeros outside the closed unit ball, and Θ_3 is a singular factor in the following sense: $\chi(\Theta_3)$ is a singular $\chi(J)$-contractive function.

Proof. The idea of the proof is to go to the complex setting using the map χ, use an analytic extension theorem, then apply Potapov's theorem and finally χ^{-1}. Since the map χ is constructed by fixing an imaginary unit $i \in \mathbb{S}$, and so a complex plane $\mathbb{C}_i$ is fixed, when we refer to the unit disc $\mathbb{D}$ we refer to the disc in $\mathbb{C}_i$. We proceed in a number of steps.

STEP 1: *The function $\chi(\Theta)$ has a meromorphic extension which is $\chi(J)$-contractive in the open unit disc $\mathbb{D}$.*

© The Author(s), under exclusive license to Springer Nature Switzerland AG 2020
D. Alpay et al., *Quaternionic de Branges Spaces and Characteristic Operator Function*, SpringerBriefs in Mathematics,
https://doi.org/10.1007/978-3-030-38312-1_7

Indeed, restricting to $p, q \in (-1, 1)$ and applying the map χ we have that the function

$$\frac{\chi(J) - \chi(\Theta)(x)\chi(J)(\chi(\Theta)(y))^*}{1 - xy} \tag{7.2}$$

is positive for x, y in $(-1, 1)$ where $\chi(\Theta)$ is defined. Let now

$$P = \frac{I + \chi(J)}{2} \quad \text{and} \quad Q = \frac{I - \chi(J)}{2}.$$

Since $\det(P + \chi(\Theta)Q) \not\equiv 0$ (to check this, reduce the situation to the case where $\chi(J)$ is replaced by a matrix of the form $\begin{pmatrix} I & 0 \\ 0 & -I \end{pmatrix}$), where I denotes the identity, one can define the Potapov-Ginzburg transform of $\chi(\Theta)$

$$\Sigma = (P + \chi(\Theta)Q)^{-1}(Q - \chi(\Theta)P).$$

It is such that:

$$\chi(J) - \chi(\Theta)(x)\chi(J)(\chi(\Theta)(y))^* = (P + \chi(\Theta)Q)\left(I - \Sigma(x)\Sigma(y)^*\right)(P + \chi(\Theta)Q)^*.$$

The kernel

$$\frac{I - \Sigma(x)\Sigma(y)^*}{1 - xy}$$

is positive definite in $(-1, 1)$. It follows that Σ is the restriction to $(-1, 1)$ of an analytic contractive function; see [4, Théorème 2.6.3, p. 44]. Going back to Θ we get that $\chi(\Theta)$ is meromorphic and J-contractive in $\mathbb{D}$.

Following [95] another possibility is to consider the function

$$V(x) = (\chi(\Theta)(x) + I)^{-1}(\chi(\Theta)(x) - I)J.$$

Then

$$\frac{V(x) + V(y)^*}{2(1 - xy)} = (\chi(\Theta)(x) + I)^{-1}\frac{J - \chi(\Theta)(x)J(\chi(\Theta)(y))^*}{1 - xy}(\chi(\Theta)(y) + I)^{-*} \tag{7.3}$$

is positive definite on $(-1, 1)$. It follows from Loewner's theorem (see [64, Theorem 1, p. 95] in the open half-plane case setting) that V is the restriction to $(-1, 1)$ of a function analytic in $\mathbb{D}$, and hence the required conclusion for $\chi(\Theta)$.

In view of Step 1, we can introduce the Potapov's decomposition of $\chi(\Theta)$,

$$\chi(\Theta)(x) = P_1(x)P_2(x)P_3(x), \tag{7.4}$$

where each of the P_u is normalized by $P_u(1) = I$, and $\chi(\Theta)(x)$ is the restriction to $(-1, 1)$ of an analytic function in the unit disc $\mathbb{D}$.

STEP 2: *There exist quaternionic Blasckhe-Potapov products of the first (resp. second) kind such that* $P_1(x) = \chi(\Theta_1(x))$ *and* $P_2(x) = \chi(\Theta_2(x))$.

The function $\chi(\Theta)$ satisfies the symmetry (4.8):

$$\chi(\Theta)(x) = \begin{pmatrix} 0 & I \\ -I & 0 \end{pmatrix} \overline{\chi(\Theta)(x)} \begin{pmatrix} 0 & -I \\ I & 0 \end{pmatrix}. \tag{7.5}$$

Also the factors in the Potapov decomposition (7.1) of $\chi(\Theta)$ are invariant under the symmetry (4.8). Since (4.8) is multiplicative (see Lemma 4.2.1), the uniqueness of the factorization implies that each of the factors in (7.1) satisfies (7.5). In the case of P_1 and P_2 we see that non-real poles or zeros will appear in pair. The same holds for P_3 when it is rational and *J*-unitary.

STEP 3: *We conclude by studying the structure of the factor* P_3:

We now study the factor $P_3(x)$, and slightly modify Potapov's original proof. In [95, pp. 216–220] Potapov proves that one can approximate uniformly on compact subsets of the open unit disc the term P_3 by rational functions. He then shows that the singularities of these rational functions converge to the unit circle $\mathbb{T}$. In fact one can directly construct such sequence of rational functions with singularities on $\mathbb{T}$. To that purpose, consider (as in (7.3) above) the function

$$V(x) = (P_3(x) + I)^{-1} (I - P_3(x)) J.$$

We have

$$\frac{V(x) + V(y)^*}{2(1 - xy)} = (P_3(x) + I)^{-1} \frac{J - P_3(x) J (P_3(y))^*}{1 - xy} (P_3(y) + I))^{-*}$$

for $x, y \in (-1, 1)$. By the already mentioned theorem of Loewner (see [64, Theorem 1, p. 95]), V has a unique analytic extension to the open unit disc, which as a real positive part there. By Herglotz's representation formula one can write in a unique way

$$V(x) = ia + \int_0^{2\pi} \frac{x + e^{it}}{x - e^{it}} dM(t), \tag{7.6}$$

the matrix a has the symmetry (7.5) and dM has the symmetry

$$\overline{dM(t)} = E \, dM(-t) \, E^{-1}$$

since the representation (7.6) is unique. Approximating dM by finite measures and so V by rational functions satisfying (7.5) we approximate $\chi(P_3)$ by finite products which satisfy (7.5) and with singularities on the unit circle, i.e., not of the form (1.15) as in [95], but of the form (4.19). Since these products converge, the limit satisfies (7.5), and we can write

$$P_3(x) = \chi(\Theta_3(x)).$$

The function $\chi(\Theta_3(x))$ is $\chi(J)$-contractive, as a limit of rational $\chi(J)$-inner functions. To finish the proof it is sufficient to take slice hyperholomorphic extension to the unit ball $\mathbb{B}$. $\square$

7.2 *J*-Contractive Functions in the Right Half-Space

We now consider the case of the right half-space which is relevant for the case of anti-self-adjoint operators.

Definition 7.2.1. *Let J be a real signature operator. The $\mathbb{H}^{n \times n}$-valued function Θ slice hyperholomorphic in an open symmetric domain is called J-contractive if the (unique) solution of the equation*

$$J - \Theta(p)J\Theta(q)^* = pK_\Theta(p,q) + K_\Theta(p,q)\overline{q} \tag{7.7}$$

is positive definite in Ω.

Proposition 7.2.2. *The $\mathbb{H}^{n \times n}$-valued function Θ is J-contractive if and only if the kernel*

$$Jk(p,q) - \Theta(p)J \star k(p,q) \star_r \Theta(q)^*, \tag{7.8}$$

where

$$k(p,q) = (\overline{p} + \overline{q})(|p|^2 + 2(\mathrm{Re}(p))\overline{q} + \overline{q}^2)^{-1},$$

is positive definite in Ω.

Proof. For $x, y \in \Omega \cap (0, \infty)$ we have

$$K_\Theta(x,y) = \frac{J - \Theta(x)J\Theta(y)^*}{x + y}.$$

Then k is left slice hyperholomorphic in p, and right slice hyperholomorphic in $\overline{q}$. Moreover $k(x,y) = \dfrac{1}{x + y}$ for $x, y \in \Omega \cap (0, \infty)$. By taking the slice hyperholomorphic extension (left) in p and (right) in $\overline{q}$ we have

$$K_\Theta(p,q) = Jk(p,q) - \Theta(p)J \star k(p,q) \star_r \Theta(q)^*. \tag{7.9}$$

                                                                                $\square$

Corollary 7.2.3. *Assume that the $\mathbb{H}^{n \times n}$-valued function Θ is J-contractive. Then $\Theta(1/p)$ is J-contractive.*

Proof. Let

$$E = \{x \in \mathbb{R} \setminus \{0\}, \text{ such that } 1/x \in \Omega\}.$$

We have:

$$K_\Theta(1/x, 1/y) = \frac{J - \Theta(1/x)J\Theta(1/y)^*}{1/x + 1/y} = x\frac{J - \Theta(1/x)J\Theta(1/y)^*}{x + y}y, \quad x, y \in E,$$

hence the kernel $\dfrac{J - \Theta(1/x)J\Theta(1/y)^*}{x + y}$ is positive definite on E. The claim follows then by slice hyperholomorphic extension. $\qquad\square$

Proposition 7.2.4. *Assume that the $\mathbb{H}^{n \times n}$-valued function Θ is slice hyperholomorphic at ∞ and that $\Theta(\infty) = I_n$. Then:*

$$\lim_{x \to \infty} F(x) = 0 \tag{7.10}$$

and the limit $\lim_{x \to \infty} x F(x)$ exists for all $F \in \mathcal{H}(\Theta)$, and defines a continuous linear operator from $\mathcal{H}(\Theta)$ into $\mathbb{H}^n$, with Riesz representation

$$d^* \left(\lim_{x \to \infty} x F(x) \right) = \langle F, g_d \rangle \tag{7.11}$$

where

$$g_d(y) = Jd - \Theta(y)Jd$$

(and in particular the function $g_d \in \mathcal{H}(\Theta)$ for every $d \in \mathbb{H}^n$).

Proof. Let $F \in \mathcal{H}(\Theta)$ and $d \in \mathbb{H}^n$. For $x \in \mathbb{R} \cap \Omega$ we have by Cauchy-Schwarz inequality

$$|d^* F(x)| = |\langle F(\cdot), K_\Theta(\cdot, x)d \rangle_{\mathcal{H}(\Theta)}|$$

$$\leq \|F\| \cdot \sqrt{d^* \frac{J - \Theta(x)J\Theta(x)^*}{2x}d} \tag{7.12}$$

$$\to 0 \quad \text{as } x \to \infty.$$

To prove the second claim we note that the norm of the operator $x \mapsto x F(x)$ is $x^2 \frac{\|J - \Theta(x)J\Theta(x)^*\|}{2x}$. In view of the analyticity at ∞, we can write

$$\Theta(x) = I_n + \frac{M}{x} + \frac{o(1/x)}{x}$$

where $\lim_{r \to \infty} o(1/x) = 0$. So

$$x\left(J - \Theta(x)J\Theta(x)^*\right) = -(MJ + JM^*) + \text{ terms which go to 0 as } x \to \infty.$$

Thus

$$\sup_{x \in \mathbb{R} \cap \Omega} x^2 \frac{\|J - \Theta(x)J\Theta(x)^*\|}{2x} < \infty.$$

It follows that for every $d \in \mathbb{H}^n$ the family of functions $xK(\cdot, x)d$ has a weak limit in $\mathcal{H}(\Theta)$. Let g_d be this limit. In a reproducing kernel Hilbert space, weak convergence implies pointwise convergence and so

$$g_d(y) = \lim_{x \to \infty} xK(y, x) = Jd - \Theta(y)Jd. \tag{7.13}$$

$\square$

Definition 7.2.5. *Assume that $\mathbb{H}^{n \times n}$-valued function Θ is slice hyperholomorphic at ∞ and that $\Theta(\infty) = I_n$. We denote by K the operator from $\mathbb{H}^n$ into $\mathcal{H}(\Theta)$ defined by*

$$(Kd)(x) = Jd - \Theta(x)Jd \tag{7.14}$$

Lemma 7.2.6. *Assume that $\mathbb{H}^{n \times n}$-valued function Θ is slice hyperholomorphic at ∞ and that $\Theta(\infty) = I_n$. Then*

$$K^*f = \lim_{x \to \infty} xf(x). \tag{7.15}$$

Proof. Let $f \in \mathcal{H}(\Theta)$ and $d \in \mathbb{H}^n$. By definition of weak limit, and using (7.13), we have:

$$\begin{aligned}
\langle K^*f, d \rangle_{\mathbb{H}^n} &= \langle f, Jd - \Theta(\cdot)Jd \rangle \\
&= \lim_{x \to \infty} x \langle F, K(\cdot, x)d \rangle \\
&= \lim_{x \to \infty} x \langle f(x), d \rangle_{\mathbb{H}^n},
\end{aligned}$$

as stated. $\square$

Theorem 7.2.7. *let J be a $\mathbb{R}^{n \times n}$-valued signature matrix. Then, the $\star$-product of J-contractive functions defined on a common axially symmetric set Ω is J-contractive.*

Proof. Let $K_1(p, q)$ and $K_2(p, q)$ be $\mathbb{H}^{n \times n}$-valued positive definite functions such that

$$J - \Theta_u(p)J\Theta_u(q)^* = pK_u(p, q) + K_u(p, q)\overline{q}, \quad u = 1, 2.$$

Then,

$$\begin{aligned}
J - (\Theta_1(p) \star \Theta_2(p))J(\Theta_1(q) \star \Theta_2(q))^* &= \\
&= J - (\Theta_1(p) \star \Theta_2(p))J(\Theta_1(q)^* \star_r \Theta_2(q)^*) \\
&= J - \Theta_1(p)J\Theta_1(q)^* + \Theta_1(p) \star \left(J - \Theta_2(p)J\Theta_2(q)^* \right) \star_r \Theta_1(q)^* \\
&= pK(p, q) + K(p, q)\overline{q}
\end{aligned}$$

with

$$K(p, q) = K_1(p, q) + \Theta_1(p) \star K_2(p, q) \star_r \Theta(q)^*.$$

This ends the proof since $K(p, q)$ is positive definite in Ω. $\square$

Theorem 7.2.8. *Let $J \in \mathbb{R}^{n \times n}$ be a signature matrix and let $A_1, \ldots, A_N \in \mathbb{H}^{n \times n}$ be such that $A_j J \geq 0$, $j = 1, \ldots, N$. Then*

$$\Theta(p) = e_\star^{-pA_1} \star \cdots \star e_\star^{-pA_N}, \tag{7.16}$$

where $e_\star^{-pA} = \sum_{n=0}^{+\infty}(-1)^n \dfrac{p^n A^n}{n!}$, is J-inner.

Proof. We first prove the claim for $N = 1$. The claim for general N follows then from Theorem 7.2.7. Set $A_1 = A$ and let $x, y \in \mathbb{R}$. Taking into account that $AJ = JA^*$, integration by part leads to:

$$
x \int_0^1 e^{-txA} A J e^{-tyA^*} dt = \left[-e^{-txA} J e^{-tyA^*} \right]_{t=0}^{t=1} + y \int_0^1 e^{-txA}(-JA^*)e^{-tyA^*} dt
$$

$$
= J - e^{-xA} J e^{-yA^*} - y \int_0^1 e^{-txA} A J e^{-tyA^*} dt,
$$

so that

$$\frac{J - e^{-xA} J e^{-yA^*}}{x + y} = \int_0^1 e^{-txA} A J e^{-tyA^*} dt, \tag{7.17}$$

and the claim follows from slice hyperholomorphic extension since $AJ \geq 0$. □

Remark 7.2.9. The preceding example is adapted from [94]. In that paper, and for complex matrices and $n = 2$ the decomposition (7.16) is shown to be unique, exhibiting a special case of the key result of de Branges [48] recalled in Theorem 1.1.4 above.

7.3 The Case of Entire Functions

The following result is partial but hints at new directions to be explored (for the uniqueness hypothesis see the book of Arov and Dym [42]).

Theorem 7.3.1. *Let Θ be a $\mathbb{H}^{n \times n}$-valued slice hyperholomorphic entire J-inner functions, where $J \in \mathbb{R}^{n \times n}$ is a real signature matrix. Then $\chi(\Theta)$ extends to a $\chi(J)$-inner entire function. If it satisfies the uniqueness representation condition, Θ can then be written as a multiplicative integral of the form*

$$\Theta(x) = \int_0^{\widetilde{\ell}} e^{xH(t)dt},$$

where H is a $\mathbb{H}^{n \times n}$-valued integrable function such that $H(t)J \geq 0$ on $[0, \ell]$.

Proof. By hypothesis, the kernel (7.9) is positive definite in the right half-space. Restricting to $p = x$ and $q = y$ in $(0, \infty)$, and applying the map χ defined by (2.2), we see that the kernel

$$\frac{\chi(J) - (\chi(\Theta)(x)) \, \chi(J) \, (\chi(\Theta)(y))^*}{x + y} \tag{7.18}$$

is positive definite in $(0, \infty)$. The rest of the proof is divided into a number of steps.

STEP1: *The power series defining $\chi(\Theta)(x)$ extends to all $\mathbb{C}$ and the corresponding kernel*

$$\frac{\chi(J) - (\chi(\Theta)(z)) \, \chi(J) \, (\chi(\Theta)(w))^*}{z + \overline{w}} \tag{7.19}$$

is positive definite in the open right half-plane.

This follows from Loewner's theorem; see Remark 1.2.3.

STEP 2: *There exist $\ell > 0$ and a $\mathbb{C}^{n \times n}$-valued function G defined on $[0, \ell]$ such that $G(t)\chi(J) \geq 0$ and*

$$\chi(\Theta)(z) = \int_0^{\ell} \widetilde{e^{zG(t)dt}},$$

where we assume $\chi(\Theta)(0) = I_n$.

Indeed, since the function $\chi(\Theta)(x)$ is real analytic in the whole real line, the function $\chi(\Theta)(z)$ is entire, and we can apply Potapov's theorem (Theorem 1.1.4 above, but for the open right half-plane rather than the open upper half-plane) to write $\chi(\Theta)$ as a multiplicative integral (1.17).

STEP 3: *We have*

$$\chi(\Theta)(x) = \int_0^{\ell} \widetilde{e^{zG_1(t)dt}}, \tag{7.20}$$

with

$$G_1(s) = \begin{pmatrix} 0 & I \\ -I & 0 \end{pmatrix} \overline{G(s)} \begin{pmatrix} 0 & -I \\ I & 0 \end{pmatrix}.$$

This follows from the limit of Θ as a limit of products of the form

$$\overset{\curvearrowright}{\prod_{k=0}^{n-1}} (I + G(s_j)(t_{j+1} - t_j))$$

from of the symmetry (7.5), and from the assumed uniqueness of the multiplicative integral representation of $\chi(\Theta)$, the function G may be chosen satisfying (7.5).

STEP 4: *We conclude the proof.*

Let $M(s, x) = \widehat{\int}_0^s e^{xG(t)dt}$. By definition of the multiplicative integral

$$\frac{d}{ds}M(s, x) = M(s, x)xG_2(s)$$

In view of (7.5) we can write

$$\frac{d}{ds}M(s, x) = M(s, x)xH(s)$$

with

$$H(s) = \frac{G_2(s) - EG_2(s)E}{2}.$$

So we can choose the matrix function in (1.17) to satisfy

$$H(s) = -EH(s)E,$$

and the result follows.

$\square$

Chapter 8
The Characteristic Operator Function

In this section we discuss the case of close-to-anti-self-adjoint operators, which is the case of interest in quaternionic analysis, and we first prove some properties of the characteristic operator function, also providing some examples. We then move to inverse problems under assumptions on the behavior at infinity, but of course other situations might be considered. We believe that this chapter can be developed in various directions and that it opens new avenues of research.

8.1 Properties of the Characteristic Operator Function

Definition 1.2.1 contains the notions of characteristic operator function in the quaternionic framework. We recall that we consider a continuous right linear operator A in a right quaternionic space, with finite dimensional real part (say of rank n) satisfying

$$A + A^* = -C^* J C, \tag{8.1}$$

where $J \in \mathbb{H}^{n \times n}$ is both self-adjoint and unitary, and C is linear bounded from the quaternionic Hilbert space $\mathcal{H}$ into $\mathbb{H}^n$. The characteristic operator function of A is defined by

$$S(p) = I_n - p C^* \star (I - pA^*)^{-\star} C J. \tag{8.2}$$

Proposition 8.1.1. *The characteristic operator function S is J-contractive.*

Proof. Define

$$K(p, q) = C^* \star (I - pA^*)^{-\star} (I - A\overline{q})^{-\star_r} \star_r C. \tag{8.3}$$

Then, $K(p, q)$ is positive definite in a neighborhood of the origin. Furthermore (see [17, Theorem 7.7])

$$J - S(p)JS(q)^* = pK(p,q) + K(p,q)\overline{q}. \tag{8.4}$$

The claim follows from Proposition 7.2.2. $\square$

When $p = q$ this equation becomes

$$J - S(p)JS(p)^* = pK(p,p) + K(p,p)\overline{p}. \tag{8.5}$$

Note that the matrix $J - S(p)JS(p)^*$ need not be nonnegative. It will be nonnegative for $x > 0$ since then we have

$$J - S(x)JS(x)^* = 2xK(x,x).$$

Example 8.1.2. We take $A = \begin{pmatrix} 0 & 1 \\ 0 & 0 \end{pmatrix}$. Then (8.1) is met with

$$C = -J = \begin{pmatrix} 0 & 1 \\ 1 & 0 \end{pmatrix}$$

and

$$S(p) = \begin{pmatrix} 1 & p \\ p & 1 + p^2 \end{pmatrix}. \tag{8.6}$$

Furthermore, for $x, y \in \mathbb{R}$ such that $x + y \neq 0$ we have:

$$\frac{J - S(x)JS(y)^*}{x + y} = \begin{pmatrix} 1 & y \\ x & 1 + xy \end{pmatrix}. \tag{8.7}$$

Example 8.1.3. We first consider the finite dimensional case. As proved in [23], the reproducing kernel Hilbert space $\mathcal{H}(\Theta)$ associated with Θ is finite dimensional if and only if Θ is rational, i.e., $\Theta(x)$ is a rational function of the real function x and with values in $\mathbb{H}^{n \times n}$.

Example 8.1.4. Here we follow [51, pp. 76–79]. Let $J \in \mathbb{H}^{n \times n}$ be a signature matrix, let $\xi(t), t \in [0, \infty)$ be a continuous $\mathbb{H}^{1 \times n}$-valued function, fix $\ell > 0$, and consider the integral operator A_ℓ defined on $\mathbf{L}_2([0, \ell], dx, \mathbb{H})$ by

$$(A_\ell f)(x) = \int_x^\ell \xi(x)J\xi(y)^* f(y)dy. \tag{8.8}$$

The operator is continuous since ξ is, and as in the complex-valued case, its adjoint is given by

$$(A_\ell^* g)(x) = \int_0^x \xi(x)J\xi(y)^* g(y)dy. \tag{8.9}$$

Indeed, with $f, g \in \mathbf{L}_2([0, \ell], dx, \mathbb{H})$, we can write:

$$\langle A_\ell f, g \rangle = \int_0^\ell g(x)^* (A_\ell f)(x) dx$$

$$= \int_0^\ell \int_0^\ell 1_{[x,\ell](y)} g(x)^* \xi(x) J \xi(y)^* f(y) dy dx$$

$$= \int_0^\ell \left(\int_0^\ell 1_{[x,\ell](y)} g(x)^* \xi(x) J \xi(y)^* dx \right) f(y) dy,$$

so that

$$(A_\ell^* g)(y) = \left(\int_0^\ell 1_{[x,\ell](y)} g(x)^* \xi(x) J \xi(y)^* dx \right)^* = \int_0^y \xi(y) J \xi(x)^* g(x) dx,$$

which is (8.9) after interchanging x and y.

We have (8.1), i.e., $A + A^* = -C^* JC$, and

$$Cf = \int_0^\ell \xi(y)^* f(y) dy.$$

Indeed

$$(A_\ell + A_\ell^*)(f) = \xi(x) J \left(\int_0^\ell \xi(y)^* f(y) dy \right).$$

The characteristic operator function of A is given by $J W(\ell, 1/p) J$ where W is the solution to the canonical differential equation

$$\frac{\mathrm{d}}{\mathrm{d}t} W(t, p) = \frac{1}{p} \xi(t)^* \xi(t) \star W(t, p). \tag{8.10}$$

Indeed, we have for real x,

$$W(t, x) = I_n + \frac{J}{x} \int_0^t \xi(u)^* \xi(u) W(u, x) du. \tag{8.11}$$

Furthermore, by definition of A_ℓ^*

$$((A_\ell^* - xI)\xi W)(t, x) = \zeta(t) \left(\int_0^t J \xi(u)^* \xi(u) W(u) du - x\xi(t) W(t) \right)$$

$$= \xi(t) \left(\int_0^t J \xi(u)^* \xi(u) W(u) du - x W(t) \right)$$

$$= x\xi(t) \left(\frac{1}{x} \int_0^t J \xi(u)^* \xi(u) W(u) du - W(t) \right)$$

$$= -x\xi(t).$$

Hence

$$\frac{1}{x}\xi(t)W(t,x) = \left((A_\ell^* - xI)^{-1}\xi\right)(t,x),$$

that is

$$\frac{\xi W}{x} = -(xI - A_\ell^*)^{-1}C.$$

Thus

$$\int_0^\ell \xi^*(t)\xi(t)W(t,x)dt = -C^*(xI - A_\ell^*)^{-1}C,$$

or, equivalently

$$J(W(\ell, x) - I_n) = -C^*(xI - A_\ell^*)^{-1}C,$$

and so

$$JW(\ell, 1/x)J = I_n - xC^*(I - xA_\ell^*)^-CJ.$$

Remark 8.1.5. Let A be a (say bounded) right linear operator from the right quaternionic Hilbert space $\mathcal{H}$ into itself, and assume that $A - A^*$ has finite rank equal to n. Then we can write

$$A - A^* = C^*EC, \tag{8.12}$$

where $E \in \mathbb{H}^{n \times n}$ satisfies

$$E^* = -E^{-1}, \tag{8.13}$$

and following (1.1), set

$$\Theta(p) = I_n - C^* \star (A - pI)^{-*}CE. \tag{8.14}$$

We have for $x, y \in \rho_S(A) \cap \mathbb{R}$

$$E - \Theta(x)E\Theta(y)^* = (x - y)C^*(A - xI)^{-1}(A - yI)^{-*}C. \tag{8.15}$$

Slice hyperholomorphic extension gives:

$$E - \Theta(p)E\Theta(q)^* = pM(p,q) - M(p,q)\overline{q}, \tag{8.16}$$

where $M(p,q)$ denotes the positive definite kernel

$$M(p,q) = C^* \star (A - pI)^{-*}(A^* - \overline{q}I)^{-*_r} \star_r C.$$

In the classical case, to consider self-adjoint or anti-self-adjoint operators is basically equivalent since they are related by a multiplication by the imaginary unit. In these two cases we have to consider, accordingly, the half-space of complex numbers with positive real part or the upper half complex plane. Here the situation is different.

Although one can consider both the cases, there is not anymore the symmetry between a real half-plane and the upper (or lower) half-plane since here we have a three dimensional imaginary space and so these notions are not meaningful. Moreover, it is not clear if Θ as in (8.16) defines an E-contractive function (even though we can say something when the function is restricted to the real axis we cannot infer its behavior on the half-space under consideration).

8.2 Inverse Problems

In the previous section we saw how to associate to any close to anti-self-adjoint operator a J-contractive function. The purpose of the present section is to study the inverse problem: given a J-contractive function, how to find, if possible, an operator whose characteristic operator function is the given J-contractive function.

Theorem 8.2.1. *Let Θ be a $\mathbb{H}^{n \times n}$-valued function, slice hyperholomorphic and J-contractive in an axially symmetric domain $\Omega \subset \mathbb{H}_+$, which is a neighborhood of infinity, and assume that $\Theta(\infty) = I_n$. Then $\Theta(1/p)$ is the characteristic operator function of a right linear continuous operator.*

Proof. The idea of the proof is to consider the reproducing kernel Hilbert space $\mathcal{H}(\Theta)$ associated with K_Θ and to build the operator in $\mathcal{H}(\Theta)$. We note that Ω contains the set $(-\infty, -x_0) \cup (x_0, \infty)$ for some $x_0 > 0$ and that, for $|x| \geq x_0$ we have

$$\Theta(-x) J \Theta(x)^* = J. \tag{8.17}$$

We proceed in a number of steps.

STEP 1: *We have that (see [25, Proof of Theorem 2.3, p. 598] for the complex case)*

$$R_a K_\Theta(x, y) = \frac{-1}{a + y} \left(K_\Theta(x, y) - K_\Theta(x, -a) J \Theta(a) J \Theta(y)^* \right). \tag{8.18}$$

Making use of (8.17) we have

$$K_\Theta(x, y) = \frac{\Theta(-y) J \Theta(y)^* - \Theta(x) J \Theta(y)^*}{x + y} = -(R_{-y}\Theta)(x) J \Theta(y)^*$$

from which (8.18) follows from the resolvent identity

$$R_a - R_b = (a - b) R_a R_b. \tag{8.19}$$

STEP 2: *Let $a, b \in \mathbb{R} \cap \Omega$. Then $R_a \mathcal{H}(\Theta) \subset \mathcal{H}(\Theta)$, R_a is bounded and*

$$\langle R_a f, g \rangle + \langle f, R_b g \rangle + (a + b) \langle R_a f, R_b g \rangle = -g(b)^* J f(a), \quad \forall f, g \in \mathcal{H}(\Theta). \tag{8.20}$$

We follow [27], where the complex setting is considered. Equality (8.20) is first proved on finite linear combinations of kernels, using (8.18). For $f = K(\cdot, y)c$ and $g = K(\cdot, z)d$, this amounts to prove:

$$
\begin{aligned}
-K(z, b)JK(a, y) = &\frac{-1}{a+y}\left(K(z, y) - K(z, -a)J\Theta(a)J\Theta(y)^*\right) - \\
&+ \frac{-1}{a+y}\left(K(z, y) - \Theta(z)J\Theta(b)^*JK(-b, y)\right) + \\
&+ \frac{a+b}{(a+y)(b+z)}\left\{\left(K(z, y) - K(z, -a)J\Theta(a)J\Theta)(y)^*\right) - \right. \\
&\left. -\Theta(z)J\Theta(b)^*JK(-b, z) + \Theta(z)J\Theta(b)^*JK(-b, -a)\Theta(a)J\Theta(y)^*\right\}.
\end{aligned}
\tag{8.21}
$$

The computation is simple but quite long and is omitted; it is maybe well to remark that, in the case $\Theta = 0$, (8.21) reduces to the easily verified equality

$$
-\frac{1}{(a+y)(x+y)} - \frac{1}{(b+z)(x+y)} + \frac{a+b}{(a+y)(b+z)(x+y)} + \frac{1}{(b+z)(a+y)} = 0.
\tag{8.22}
$$

Since point evaluations are bounded, there exists $C_a > 0$ such that

$$
\|f(a)\| \le C_a \cdot \|f\|, \quad f \in \mathcal{H}(\Theta).
\tag{8.23}
$$

Setting $a = b$ we get then from (8.20) the inequality of the form

$$
\|R_a f\|^2 \le \frac{1}{2a}\left(2\|R_a f\| \cdot \|f\| + 2C_a^2\|f\|^2\right),
\tag{8.24}
$$

for f which are linear combinations of functions of the form $K(\cdot, x)\xi$, where $x \in \Omega \cap \mathbb{R}$ and $\xi \in \mathbb{H}^n$. It follows that

$$
\frac{\|R_a f\|^2}{\|f\|^2} \le \frac{\|R_a f\|}{a\|f\|} + C_a^2
$$

for such f. It follows that R_a has a bounded extension to all of $\mathcal{H}(\Theta)$.

STEP 3: $\ker R_a = \{0\}$ for $a \in \mathbb{R}_+ \cap \Omega$, and in particular for $f \in \mathcal{H}(\Theta)$ there exists at most one $c_f \in \mathbb{H}^n$ such that the function

$$
p \star f(p) + c_f
\tag{8.25}
$$

belongs to $\mathcal{H}(\Theta)$.

By the resolvent identity (8.19), we have $\ker R_a = \ker R_b$ for any choice of $a, b \in \mathbb{R} \cap \Omega$. Let $\xi \ne 0 \in \ker R_a$ and let $F \in \mathcal{H}(\Theta)$. Then, by Proposition 7.2.4 we have $\lim_{x \to \infty} F(x) = 0$, contradicting $\xi \ne 0$ when $\xi \in \mathcal{H}(\Theta)$.

STEP 4: *The operator A defined by (8.25) and with domain*

Dom $A = \{f \in \mathcal{H}(\Theta) : \exists c_f \in \mathbb{H}^n$ for which the function (8.25) belongs to $\mathcal{H}(\Theta)\}$

is closed.

Functions whose restriction to $\mathbb{R} \cap \Omega$ is of the form

$$f(x) = \sum_{u=1}^{U} K_\Theta(x, x_u)\xi_u \quad \text{with} \quad \sum_{u=1}^{U} \Theta(x_u)^*\xi_u = 0$$

belong to the domain of A since

$$xf(x) = \sum_{u=1}^{U} x \frac{J - \Theta(x)J\Theta(x_u)^*}{x + x_u} \xi_u$$

$$= \sum_{u=1}^{U} \frac{x + x_u - x_u}{x + x_u} \left(J - \Theta(x)J\Theta(x_u)^*\right) \xi_u$$

$$= -\sum_{u=1}^{U} x_u \frac{J - \Theta(x)J\Theta(x_u)^*}{x + x_u} \xi_u + J \left(\sum_{u=1}^{U} \xi_u\right) -$$

$$- \Theta(x)J \left(\sum_{u=1}^{U} \Theta(x_u)^*\xi_u\right).$$

It follows that the domain of A is dense. Indeed, let $f \in \mathcal{H}(\Theta)$ orthogonal to the domain of A. In particular

$$\langle f, K_\Theta(\cdot, x)\xi - K(\cdot, y)\Theta(y)^{-*}\Theta(x)^*\xi \rangle_{\mathcal{H}(\Theta)} = 0$$

for all $x, y \in \mathbb{R} \cap \Omega$ and $\xi \in \mathbb{H}^n$. Thus, for all such x, y and ξ we have

$$\xi^* f(x) \equiv \xi^* \Theta(x)\Theta(y)^{-1} f(y)$$

and so the function $x \mapsto \Theta(x)^{-1} f(x)$ is constant, i.e., $f(x) = \Theta(x)\eta$ for some $\eta \in \mathbb{H}^n$. Letting $x \longrightarrow \infty$, and since $\Theta(\infty) = I_n$, we get a contradiction with (7.12). Let now $a \in \mathbb{R} \cap \Omega$. Since A is one-to-one and densely defined, the operator $(A - aI)^{-1}$ exists as a possibly unbounded densely defined operator. Let $f \in \text{Ran}\,(A - aI)$. Then $f = (A - aI)g$ for some $g \in \mathcal{H}(\Theta)$. Thus for $p = x \in \mathbb{R} \cap \Omega$ we have

$$f(x) = xg(x) + c_g - ag(x),$$

so that

$$g(x) = \frac{f(x) + c_g}{x - a}. \tag{8.26}$$

Since g is continuous across $\mathbb{R} \cap \Omega$ we have $c_g = -f(a)$ and $g = R_a f$. Since we proved that R_a is continuous we have $\mathbb{R} \cap \Omega \subset \rho_S(A)$. Here the S-spectrum reduces to the classical spectrum since a is real.

Take now $a = -b$ in (8.20) and $f = (A - a)F$ and $g = (A + a)G$. Then

$$f(a) = c_F \quad \text{and} \quad g(a) = c_G,$$

and (8.20) becomes (with K defined by (7.2.6))

$$\langle F, AG \rangle + \langle AF, G \rangle = -\langle f, KJK^*g \rangle.$$

It follows from this identity that A has an adjoint and is a bounded operator, and that, moreover,

$$A + A^* = -K'JK^*. \tag{8.27}$$

We can now conclude the proof.

Let $d \in \mathbb{N}$ and $x \in \mathbb{R} \cap \Omega$. We write

$$K^*(T - xI)^{-1}KJd = K^*(T - x)^{-1}(J - \Theta(\cdot)J)Jd = K^*G,$$

where

$$G(t) = \frac{J - \Theta(t)J - (J - \Theta(x)J)}{t - x}Jd = \frac{\Theta(x) - \Theta(t)}{t - x}d.$$

Computing

$$K^*G = \lim_{t \to \infty} tG(t) = (\Theta(x) - I_n)d.$$

Thus

$$\Theta(x)d = d + K^*G = d + K^*(T - xI)^{-1}KJ$$

and so

$$\Theta(x) = I_n - K^*(xI - A)^1 KJ \quad \text{and} \quad \Theta(1/x) = I_n - xK^*(xI - A)^{-1}KJ \tag{8.28}$$

and so Θ is the characteristic operator function of A^*.   □

Chapter 9
$\mathcal{L}(\Phi)$ Spaces and Linear Fractional Transformations

In the complex setting case (and in the related time-varying case), spaces $\mathcal{L}(\Phi)$ and $\mathcal{H}(\Theta)$ (and the related $\mathcal{H}(A, B)$ spaces) are closely related by linear fractional transformations; this originates with the works of de Branges, see [47] and of de Branges and Rovnyak [50]. In this section we outline the counterpart of such relations in the quaternionic setting. We work in the framework of the slice hyperholomorphic weights, introduced in Section 3.4 and we also need the definition of Carathéodory and Herglotz functions.

Definition 9.0.2. *A $\mathbb{H}^{n \times n}$-valued function, say Φ, slice hyperholomorphic in an axially symmetric open Ω subset of the right half-space and such that the kernel*

$$K_\Phi(x, y) = \frac{\Phi(x) + \Phi(y)^*}{x + y}$$

is positive definite in $\Omega \cap \mathbb{R}$ is called a Carathéodory function.
A $\mathbb{H}^{n \times n}$-valued functions, say Φ, slice hyperholomorphic in an axially symmetric open Ω subset of the unit ball, respectively, and such that the kernel

$$K_\Phi(x, y) = \frac{\Phi(x) + \Phi(y)^*}{1 - xy}$$

is positive definite in $\Omega \cap \mathbb{R}$ is called a Herglotz function.

9.1 $\mathcal{L}(\Phi)$ Spaces Associated with Analytic Weights

We focus on the case of Carathéodory functions, and will only mention briefly some of the results pertaining to Herglotz functions.

D. Alpay et al., *Quaternionic de Branges Spaces and Characteristic Operator Function*, SpringerBriefs in Mathematics, https://doi.org/10.1007/978-3-030-38312-1_9

Theorem 9.1.1. *Let W_+ be a $\mathbb{H}^{n \times n}$-valued slice hyperholomorphic weight in an axially symmetric domain $\Omega \subseteq \mathbb{H}$ such that $\int_{\mathbb{R}} \|W_+(it)\|^2 dt < \infty$. The function*

$$\Phi(a) = \int_{\mathbb{R}} W_+(it)^* (W_+(p) \star (a - p)^{-*})_{|p=it} dt \qquad (9.1)$$

extends uniquely to the right half-space to a Carathéodory function, still denoted by Φ.

Proof. For $a, b \in \Omega \cap \mathbb{R}$ we have to compute

$$\Phi(a) + \Phi(b)^*$$
$$= \int_{\mathbb{R}} W_+(it)^* [W_+(p) \star (a - p)^{-*}]_{|p=it} \, dt + \int_{\mathbb{R}} [(b + \bar{p})^{-*_r} \star_r W_+(p)^*]_{|p=it} W_+(it) \, dt$$
$$= \int_{\mathbb{R}} W_+(it)^* [(a - p)^{-1} W_+(p)]_{|p=it} \, dt + \int_{\mathbb{R}} [W_+(p)^* (b + \bar{p})^{-1}]_{|p=it} W_+(it) \, dt$$
$$= \int_{\mathbb{R}} W_+(it)^* (a - it)^{-1} W_+(it) \, dt + \int_{\mathbb{R}} W_+(it)^* (b - it)^{-1} W_+(it) \, dt$$
$$= \int_{\mathbb{R}} W_+(it)^* [(a - it)^{-1} + (b + it)^{-1}] W_+(it) \, dt$$
$$= (a + b) \int_{\mathbb{R}} W_+(it)^* [(a - it)^{-1} (b + it)^{-1}] W_+(it) \, dt$$

which expresses that $\dfrac{\Phi(a) + \Phi(b)^*}{a + b}$ is positive definite in $\Omega \cap \mathbb{R}$. The result follows by slice hyperholomorphic extension. $\qquad\square$

Theorem 9.1.2. *The associated reproducing kernel Hilbert space is given by the closure of the set of functions of the form*

$$F(a) = \int_{\mathbb{R}} W_+(it)^* \left(f(p) \star (a - p)^{-*} \star W_+(p) \right)_{|p=it} dt \qquad (9.2)$$

where f belongs to the right linear span of the functions $p \mapsto (p + a)^{-*}$, and with norm

$$\|F\|_{W_+}^2 = \int_{\mathbb{R}} \|(f(p) \star W_+(p))_{|p=it}\|^2 dt.$$

Proof. Let $N \in \mathbb{N}$, $x_1, \ldots, x_N \in \Omega \cap \mathbb{R}$ and $c_1, \ldots, c_N \in \mathbb{H}^n$. The claim follows from the formula

$$\sum_{n=1}^{N} K_\Phi(x, x_n) c_n = \int_{\mathbb{R}} W_+(it)^* \left(f(p) \star (a - p)^{-*} \star W_+(p) \right)_{|p=it} dt,$$

where $f(p) = \sum_{n=1}^{N} (x_n + p)^{-*} c_n$ is such that

$$\|\sum_{n=1}^{N} K_\Phi(\cdot, x_n)c_n\|_{W_+}^2 = \int_{\mathbb{R}} \|((f \star W_+)(p))_{|p=it}\|^2 dt. \qquad \square$$

Proposition 9.1.3. *The space $\mathcal{L}(\Phi)$ is R_b-invariant for $b \in \Omega \cap \mathbb{R}$. Furthermore R_b is bounded and it holds that*

$$\langle R_a F, G \rangle_{W_+} + \langle F, R_b G \rangle_{W_+} + (a+b)\langle R_a F, R_b G \rangle_{W_+} = 0, \quad \forall F, G \in \mathcal{L}(\Phi), \tag{9.3}$$

Proof. Let F be of the form (9.2). Letting $a \in \mathbb{R}$ going to infinity shows that the only constant function is 0. Furthermore we have

$$\frac{F(a) - F(b)}{a - b} = \int_{\mathbb{R}} W_+(it)^* \left(f(p) \star (b-p)^{-\star}\right) \star (a-p)^{-\star})_{|p=it} dt$$

and so

$$\|R_b F\|^2 = \int_{\mathbb{R}} \|(f(p) \star (b-p)^{-\star})_{|p=it} W_+(it)\|^2 dt$$

$$= \int_{\mathbb{R}} \frac{\|(f(p) \star (b-p)^{-\star})_{|p=it} W_+(it)\|^2}{b^2 + t^2} dt \leq \frac{1}{b^2}\|F\|^2.$$

Equality (9.3) follows from (8.22). $\qquad \square$

Proposition 9.1.4. *The space $\mathcal{L}(\Phi)$ contains no nonzero constant element. To see that, let $a \to +\infty$ in (9.2). Hence the operator*

$$M_p F = pF + c_F$$

where $c_F \in \mathbb{H}^n$ is uniquely defined, is anti-self-adjoint in $\mathcal{L}(\Phi)$ and satisfies

$$R_b = (M_p - bI)^{-1}, \quad b \in \Omega \cap \mathbb{R}. \tag{9.4}$$

Proof. Let F be of the form (9.2). We have

$$aF(a) = a \int_{\mathbb{R}} W_+(it)^* \left(f(p) \star (a-p)^{-\star} \star W_+(p)\right)_{|p=it} dt$$

$$= \int_{\mathbb{R}} W_+(it)^* \left(f(p) \star (a-p+p) \star (a-p)^{-\star} \star W_+(p)\right)_{|p=it} dt$$

$$= \int_{\mathbb{R}} W_+(it)^* \left(f(p) \star (p) \star (a-p)^{-\star} \star W_+(p)\right)_{|p=it} dt$$

$$+ \underbrace{\int_{\mathbb{R}} W_+(it)^* (f(p) \star W_+(p))_{|p=it} dt}_{-c_F},$$

and so M_p is unitarily equivalent to multiplication by it in $\mathbf{L}_2(W_+^* W_+, dt)$, and hence anti-self-adjoint, since the above equality extends by slice hyperholomorphic extension. Furthermore, let F and G in $\mathcal{L}(\Phi)$ be such that $(\mathsf{M}_p - bI)^{-1} F = G$. Then, for real $a \in \Omega$ we have

$$F(a) = aG(a) + c_G - bG(a),$$

so that

$$G(a) = \frac{F(a) - c_G}{a - b}.$$

It follows that $c_G = F(a)$ and that (9.4) holds.   □

We now turn to the case of the quaternionic unit ball, namely, to the case of Herglotz functions. We define

$$\Phi(x) = \int_0^{2\pi} W_+(e^{it})^* \left(\left(\frac{p+x}{p-x} \right) \star W_+(p) \right)_{|p=e^{it}} dt, \quad x \in (-1, 1). \qquad (9.5)$$

The proof of the following proposition is similar to that of Proposition 9.1.3 and is omitted.

Proposition 9.1.5. *The function $\Phi(x)$ extends to a uniquely defined Herglotz function in $\mathbb{B}_1$. The associated reproducing kernel Hilbert space $\mathcal{L}(\Phi)$ is R_b-invariant for $b \in (-1, 1)$. The operators R_b are bounded in $\mathcal{L}(\Phi)$ and satisfy*

$$\langle F, G \rangle + a \langle R_a F, G \rangle + b \langle F, R_b G \rangle + (1 - ab) \langle R_a F, R_b G \rangle = 0 \qquad (9.6)$$

where $F, G \in \mathcal{L}(\Phi)$ and $a, b \in (-1, 1)$.

9.2 Linear Fractional Transformations and an Inverse Problem

As in the complex setting case, and as one can see from our previous works [1, 8, 23], linear fractional transformations play a key role in Schur analysis and related problems in operator theory. The following result was first proved by de Branges and Rovnyak in [49]. The argument is essentially the same.

Theorem 9.2.1. *Let J_1 be given by (1.7), and let Θ and Φ be, respectively, J_1-contractive and of Carathéodory class, with corresponding reproducing kernel Hilbert spaces $\mathcal{H}(\Theta)$ and $\mathcal{L}(\Phi)$. Then the map τ*

$$\tau: \quad F \mapsto \left(\Phi \ I_n \right) \star F$$

is a contraction from $\mathcal{H}(\Theta)$ into $\mathcal{L}(\Phi)$ if and only if there is a Schur multiplier S such that

$$\Phi \star (\Theta_{11} - \Theta_{12} - (\Theta_{11} + \Theta_{12}) \star S) = (\Theta_{21} + \Theta_{22}) \star S + (\Theta_{22} - \Theta_{21}) \quad (9.7)$$

Proof. Assuming τ contractive (and hence bounded). As we know from [23], its adjoint is given by

$$\tau^* (K_\Phi(\cdot, b)c) = K_\Theta(\cdot, b) \star_r \begin{pmatrix} \Phi(b)^* c \\ c \end{pmatrix}.$$

The fact that τ is contractive is equivalent to the fact that the kernel

$$K_\Phi(p, q) - (\Phi \ I_n) \star K_\Theta(p, q) \star_r \begin{pmatrix} \Phi(q)^* c \\ c \end{pmatrix}$$

is positive definite in Ω. To finish the proof note that this kernel can be rewritten as

$$(\Phi(x) \ I_n) \frac{\Theta(x) J_1 \Theta(y)^*}{x + y} \begin{pmatrix} \Phi(y)^* c \\ c \end{pmatrix}$$

is positive definite for $x, y \in \Omega \cap \mathbb{R}$. This kernel in turn can be written as

$$\frac{(\Phi\Theta_{11} + \Theta_{21})(\Phi\Theta_{12} + \Theta_{22})^* + (\Phi\Theta_{12} + \Theta_{22})(\Phi\Theta_{11} + \Theta_{21})}{x + y} =$$

$$= \frac{(\Phi\Theta_{11} + \Theta_{21} + \Phi\Theta_{12} + \Theta_{22})(\Phi\Theta_{11} + \Theta_{21} + \Phi\Theta_{12} + \Theta_{22})^*}{2(x + y)}$$

$$- \frac{(\Phi\Theta_{11} + \Theta_{21} - \Phi\Theta_{12} - \Theta_{22})(\Phi\Theta_{11} + \Theta_{21} - \Phi\Theta_{12} - \Theta_{22})^*}{2(x + y)}.$$

To conclude one uses the slice hyperholomorphic version of Leech's theorem; see [23, Section 11.6]. To prove the converse just read backwards. $\square$

In the complex setting, the question of finding all the linear fractional representations of a given Carathéodory (or Herglotz) function allows to gather under a common framework a wide range of problems, ranging from interpolation problems to inverse spectral problems. It bears the name *lossless inverse scattering problem* because of its connections with linear network theory; see Definition 1.1.1. See [4, 62, 63, 96] for the links with network theory. A systematic study of this problem was done in [25, 26]. The following result follows [25, Theorem 3.1, p. 600] and gives a family of solutions to the corresponding problem in the quaternionic setting. Recall that the space $\mathcal{M}^\square$ was introduced in the proof of Theorem 3.4.3. See Definition 3.4.4.

Theorem 9.2.2. *Let W_+ be an analytic weight and let M be a resolvent invariant subspace of* $\mathbf{L}_2(W_+^* W_+)$*, with extension $\mathcal{M}^\square$. The map $F \mapsto (\Phi \ I_n) F$ is an isometry from $\mathcal{M}^\square$ into $\mathcal{L}(\Phi)$.*

Proof. By definition of f_- we have

$$\left(\Phi \; I_n\right) F(a) = \Phi(a) f(a) + f_-(a)$$

$$= \int_{\mathbb{R}} W_+(it)^* [W_+(p) \star \{(p-a)^{-*} \star (f(p) - f(a)) + (p-a)^{-*} f(a)\}]_{p=it}\, dt$$

$$= \int_{\mathbb{R}} W_+(it)^* [W_+(p) \star f(p) \star (p-a)^{-*}]_{p=it}\, dt,$$

so the map is an isometry. $\square$

It follows from Theorem 3.4.3 that $\mathcal{M}^{\square}$ is a $\mathcal{H}(\Theta)$ space and from Theorem 9.2.1 that Φ and Θ are related by a linear fractional transformation.

We can now present the counterpart of Definition 1.1.1. In the classical case, this definition has connections with numerous questions in analysis and operator theory such as interpolation problems and models for pairs of anti-self-adjoint operators.

Definition 9.2.3. *For a given analytic weight W_+ we will call lossless inverse scattering problem the question of describing the associated function Φ as a linear fractional transformation of the form (9.2.1).*

Chapter 10
Canonical Differential Systems

Canonical differential systems and their connections to operators have a long history; see, e.g., [2, 41, 69, 103, 104]. In this section we consider such systems in the quaternionic setting, in particular in the case of rational spectral data. We foresee that these systems have potentially several applications, for example, to non-linear partial differential equations and inverse scattering.

10.1 The Matrizant

Let J be a signature matrix with real entries. Canonical differential systems are differential equations of the form (8.10), or of the more particular form

$$J \frac{d}{dt} F(t, p) = (pI + H(t)) \star F(t, p),$$ (10.1)

where p is a quaternionic variable. These systems have been thoroughly studied in the complex setting, see [102], since they provide a convenient unifying framework to discuss a number of questions pertaining to inverse scattering, non-linear partial differential equations, and other topics. The solution of either of these equations subject to the initial condition I_n is called the *matrizant*.

Of particular interest is the case where $J = J_0$ (with J_0 as in (1.20)) and H is of the form

$$H(t) = \begin{pmatrix} 0 & v(t) \\ -v(t)^* & 0 \end{pmatrix},$$

that is

$$J_0 \frac{d}{dt} F(t, p) = \left(pI + \begin{pmatrix} 0 & v(t) \\ -v(t)^* & 0 \end{pmatrix} \right) \star F(t, p).$$ (10.2)

© The Author(s), under exclusive license to Springer Nature Switzerland AG 2020
D. Alpay et al., *Quaternionic de Branges Spaces and Characteristic
Operator Function*, SpringerBriefs in Mathematics,
https://doi.org/10.1007/978-3-030-38312-1_10

Proposition 10.1.1. *The matrizant is an entire function of p.*

Proof. It suffices to apply the map ω (see (3.18)) to (10.2), and use the corresponding complex-variable result (see, e.g., [69] for the latter).   □

In [35] a class of potentials, called pseudo-exponential potentials, was introduced; these potentials correspond to rational characteristic spectral functions, and in the present setting are given by the formula

$$v(t) = -2ce^{-2ta} \left(I_m + \Omega Y - \Omega e^{-2ta^*} Y e^{-2ta}\right)^{-1} (b + \Omega c^*). \tag{10.3}$$

In this expression, $(a, b, c) \in \mathbb{H}^{m \times m} \times \mathbb{H}^{m \times n} \times \mathbb{H}^{n \times m}$, with the property that the S-spectra of a and of $a^\times = a - bc$ are in the right half-space and Ω and Y belong to $\mathbb{H}^{m \times m}$ and are the unique solutions of the Lyapunov equations

$$Ya + a^*Y = c^*c \tag{10.4}$$

$$\Omega a^{\times *} + a^\times \Omega = bb^*. \tag{10.5}$$

Because of the hypothesis on the spectra of a and $a^\times$ we note that Ω and Y are strictly positive and thus $(I_m + Y\Omega)$ is invertible. For future reference we note that

$$\Omega(I_m + Y\Omega)^{-1} = \sqrt{\Omega}(I_m + \sqrt{\Omega}Y\sqrt{\Omega})^{-1}\sqrt{\Omega} > 0, \tag{10.6}$$

where $\sqrt{\Omega}$ denotes the positive squareroot of Ω.

An explicit formula for the matrizant is given in the following theorem; see [37, Theorem 2.1, p. 9]; we follow the notation of that paper, but the reader should bear in mind differences in signs and variables because of the present setting. In a way similar to [37] we set $f = (b^* + c\Omega)(I_m + Y\Omega)^{-1}$,

$$F = \begin{pmatrix} c & 0 \\ 0 & f \end{pmatrix}, \quad T = \begin{pmatrix} -a & 0 \\ 0 & -a^* \end{pmatrix},$$

$$G = \begin{pmatrix} 0 & -f^* \\ -c^* & 0 \end{pmatrix}, \quad Z = \begin{pmatrix} 0 & \Omega(I_m + Y\Omega)^{-1} \\ Y & 0 \end{pmatrix},$$

and

$$Q(t, s) = Fe^{tT} \left(I_{2m} - e^{tT} Z e^{tT}\right)^{-1} e^{sT} G. \tag{10.7}$$

The invertibility of $X(t) = I_{2m} - e^{tT} Z e^{tT}$ for $t > 0$ will be proved in Theorem 10.1.2. Furthermore we note that

$$T J_0 = J_0 T, \quad F J_0 = J_0 F, \quad G J_0 = -J_0 G, \quad \text{and} \quad Z J_0 = -J_0 Z. \tag{10.8}$$

To ease the notation, we do not specify the size of the identity matrices in J_0. These will always be understood from the context.

Theorem 10.1.2. *Let J_0 be as in (1.20) and let Q be given by (10.7). Then, the function*

$$U(t,q) = e^{tqJ_0} + \int_t^\infty Q(t,s) \star e^{sqJ_0} ds \qquad (10.9)$$

is the unique solution to the canonical system (10.2) subject to the boundary condition

$$\lim_{\substack{q=x\in\mathbb{R} \\ t\to\infty}} e^{-txJ_0} U(t,x) = I_{2n}, \qquad (10.10)$$

where the potential $v(t)$ in (10.2) is given by (10.3).

Remark 10.1.3. In the framework of complex numbers, the factor e^{-txJ_0} is replaced by e^{-itxJ_0}, allowing an interpretation in terms of incoming and outgoing wave, hence the terminology terms such as scattering function and the like. Here, as we already noticed, the boundary is made of the purely imaginary quaternions. We do not have a wave interpretation of the various functions associated in Section 10.2 to U.

Proof of Theorem 10.1.2. We follow the arguments of [37, Proof of Theorem 2.1, p. 9], and proceed in a number of steps.

STEP 1: *Z is the (unique) solution of the Lyapunov equation*

$$TZ + ZT = GF \qquad (10.11)$$

Indeed,

$$TZ + ZT = -\begin{pmatrix} 0 & a\Omega(I_m + Y\Omega)^{-1} + \Omega(I_m + Y\Omega)^{-1}a^* \\ a^*Y + Ya & 0 \end{pmatrix}.$$

The $(2,1)$ block is equal to c^*c in view of (10.4). Multiplying the $(1,2)$ block by (-1) we can write (using (10.4) and (10.5) to go from the third line to the fifth one)

$$a\Omega(I_m + Y\Omega)^{-1} + (I_m + \Omega Y)^{-1}\Omega a^* =$$
$$= (I_m + \Omega Y)^{-1}\left(a^*(I_m + Y\Omega) + (I_m + \Omega Y)a\Omega\right)(I_m + Y\Omega)^{-1}$$
$$= (I_m + \Omega Y)^{-1}\left(\Omega a^* + a\Omega + \Omega(a^*Y + Ya)\Omega\right)(I_m + Y\Omega)^{-1}$$
$$= (I_m + \Omega Y)^{-1}\left(\Omega a^{\times\times} + a^\times\Omega + \Omega c^*b^* + bc\Omega + \Omega(a^*Y + Ya)\Omega\right)(I_m + Y\Omega)^{-1}$$
$$= (I_m + \Omega Y)^{-1}\left(bb^* + \Omega c^*b^* + bc\Omega + \Omega c^*c\Omega\right)(I_m + Y\Omega)^{-1}$$
$$= (I_m + \Omega Y)^{-1}\left((b + \Omega c^*)(b + \Omega c^*)^*\right)(I_m + Y\Omega)^{-1}$$
$$= f^*f,$$

and hence the result since

$$GF = \begin{pmatrix} 0 & -f^*f \\ -c^*c & 0 \end{pmatrix}.$$

STEP 2: *The matrix function* $X(t) = I_{2m} - e^{tT}Ze^{tT}$ *is invertible for every real* t.
To prove the claim we write

$$X(t) = \begin{pmatrix} I_m & -e^{-ta}\Omega(I_m + Y\Omega)^{-1}e^{-ta^*} \\ -e^{-ta^*}Ye^{-ta} & I_m \end{pmatrix}$$

By Schur complement $X(t)$ is invertible if and only if

$$I_m + e^{-ta}\Omega(I_m + Y\Omega)^{-1}e^{-ta^*}e^{-ta^*}Ye^{-ta}$$

is invertible. But the matrices

$$e^{-ta}\Omega(I_m + Y\Omega)^{-1}e^{-ta^*} \quad \text{and} \quad e^{-ta^*}Ye^{-ta}$$

are positive (in fact strictly positive since Y and $\Omega(I_m + Y\Omega)^{-1}$ are strictly positive; see (10.6) for the latter). So $X(t)$ is invertible. In view of later computation we note that

$$X^{-1}(t) = \begin{pmatrix} I_m & -e^{-ta}\Omega(I_m + Y\Omega)^{-1}e^{-ta^*} \\ e^{-ta^*}Ye^{-ta} & I_m \end{pmatrix} \cdot \begin{pmatrix} A_{11} & 0 \\ 0 & A_{22} \end{pmatrix}, \quad (10.12)$$

where

$$A_{11} := (I_m + e^{-ta}\Omega(I_m + Y\Omega)^{-1}e^{-2ta^*}Ye^{-ta})^{-1},$$

and

$$A_{22} := (I_m + e^{-ta^*}Ye^{-2ta}\Omega(I_m + Y\Omega)^{-1}e^{-ta^*})^{-1}.$$

STEP 3: $Q(t, s)$ *satisfies the differential equation*

$$J_0\frac{\partial Q}{\partial t} + \frac{\partial Q}{\partial s}J_0 = (J_0Q(t, t) - Q(t, t)J_0)Q(t, s). \quad (10.13)$$

With $X(t) = I_{2m} - e^{tT}Ze^{tT}$ for $t > 0$, and using the Lyapunov equation (10.11), we have

$$X'(t) = -e^{tT}TZe^{tT} - e^{tT}ZTe^{tT} = -e^{tT}GFe^{tT}.$$

Hence we can write:

$$\begin{aligned}
\frac{\partial Q}{\partial t} &= Fe^{Tt}X^{-1}e^{sT}G - Fe^{tT}X^{-1}X'X^{-1}e^{st}G \\
&= Fe^{Tt}TX^{-1}e^{sT}G + Fe^{tT}X^{-1}e^{tT}GFe^{tT}X^{-1}e^{st}G \\
&= Fe^{Tt}TX^{-1}e^{sT}G + Q(t, t)Q(t, s).
\end{aligned}$$

On the other hand,

$$\frac{\partial Q}{\partial s} = F e^{Tt} X^{-1} e^{sT} T G.$$

Hence, using (10.8),

$$
\begin{aligned}
J_0 \frac{\partial Q}{\partial t} + \frac{\partial Q}{\partial s} J_0 &= F e^{Tt} T J_0 X^{-1} e^{sT} G + J_0 Q(t,t) Q(t,s) - F e^{Tt} X^{-1} J_0 e^{sT} T G \\
&= J_0 Q(t,t) Q(t,s) + F e^{tT} X^{-1} (-J_0 T X + X T J_0) X^{-1} e^{sT} G \\
&= J_0 Q(t,t) Q(t,s) + F e^{tT} X^{-1} e^{tT} (T J_0 Z - Z T J_0) e^{tT} X^{-1} e^{sT} G \\
&= J_0 Q(t,t) Q(t,s) - F e^{tT} X^{-1} e^{tT} (T Z J_0 + Z T J_0) e^{tT} X^{-1} e^{sT} G \\
&= J_0 Q(t,t) Q(t,s) - F e^{tT} X^{-1} e^{tT} G F J_0 e^{tT} X^{-1} e^{sT} G \\
&= (J_0 Q(t,t) - Q(t,t) J_0) Q(t,s).
\end{aligned}
$$

STEP 4: *We have*

$$J_0 Q(t,t) - Q(t,t) J_0 = \begin{pmatrix} 0 & v(t) \\ v(t)^* & 0 \end{pmatrix}$$

with v given by (10.3).

Let $(Q_{ij})_{i,j=1,2}$ denote the block matrix decomposition of Q. We first note that

$$J_0 Q(t,t) - Q(t,t) J_0 = \begin{pmatrix} 0 & 2Q_{12}(t,t) \\ -2Q_{21}(t,t) & 0 \end{pmatrix}.$$

We have

$$\left(I_{2m} - e^{tT} Z e^{tT}\right) = \begin{pmatrix} I_m & -e^{-ta} \Omega (I_m + Y\Omega)^{-1} e^{-ta^*} \\ -e^{-ta^*} Y e^{-ta} & I_m \end{pmatrix},$$

and so

$$\left(I_{2m} - e^{tT} Z e^{tT}\right)^{-1} = \begin{pmatrix} I_m & e^{-ta} \Omega (I_m + Y\Omega)^{-1} e^{-ta^*} \\ e^{-ta^*} Y e^{-ta} & I_m \end{pmatrix} \begin{pmatrix} \Delta_1^{-1} & 0 \\ 0 & \Delta_1^{-*} \end{pmatrix}$$

with

$$\Delta_1 = I_m - e^{-ta} \Omega (I_m + Y\Omega)^{-1} e^{-2ta^*} Y e^{-ta}.$$

Hence

$$
Q(t,t) = - \begin{pmatrix} c e^{-ta} & 0 \\ 0 & f e^{-ta^*} \end{pmatrix} \begin{pmatrix} I_m & e^{-ta} \Omega (I_m + Y\Omega)^{-1} e^{-ta^*} \\ e^{-ta^*} Y e^{-ta} & I_m \end{pmatrix} \times
$$
$$
\times \begin{pmatrix} \Delta_1^{-1} & 0 \\ 0 & \Delta_1^{-*} \end{pmatrix} \begin{pmatrix} 0 & e^{-ta} f^* \\ e^{-ta^*} c^* & 0 \end{pmatrix},
$$

so that

$$Q_{12}(t) = Q_{21}(t)^* = -ce^{-ta}\Delta_1^{-1}e^{-ta}f^*.$$

To conclude we recall that $f = (b^* + c\Omega)(I_m + Y\Omega)^{-1}$; since

$$\Omega(I_m + Y\Omega)^{-1} = (I_m + \Omega Y)^{-1}\Omega$$

we can write

$$-ce^{-ta}\Delta_1^{-1}e^{-ta}f^*$$
$$= -ce^{-2ta}(I_m + \Omega(I_m - Y\Omega)^{-1}e^{-2ta^*}Ye^{-2ta})^{-1}(I_m + \Omega Y)^{-1}(b + \Omega c^*)$$
$$= -ce^{-2ta}(I_m + \Omega Y - \Omega e^{-2ta^*}Ye^{-2ta})^{-1}(b + \Omega c^*).$$

We note that the matrix function

$$I_m + \Omega Y - \Omega e^{-2ta^*}Ye^{-2ta} = \sqrt{\Omega}\left(I_m + \sqrt{\Omega}(Y - \Omega e^{-2ta^*}Ye^{-2ta})\sqrt{\Omega}\right)\sqrt{\Omega}^{-1}$$

is invertible for all $t \geq 0$ since $Y \geq e^{-2ta^*}Ye^{-2ta}$ for all such $t \geq 0$.

STEP 5: *The function U defined by* (10.9) *is a solution of the canonical system* (10.2).

We first take $p = x$ real. We can write:

$$J_0\frac{d}{dt}U(t, x) = xe^{txJ_0}I_n - J_0Q(t, t)e^{txJ_0} + \int_t^\infty J_0\frac{\partial Q}{\partial t}(t, s)e^{sxJ_0}ds$$
$$= xe^{txJ_0}I_{2n} - J_0Q(t, t)e^{txJ_0} +$$
$$\quad + \int_t^\infty \left(-\frac{\partial Q}{\partial s}(t, s)J_0 + (J_0Q(t, t) - Q(t, t)J_0)Q(t, s)\right)e^{sxJ_0}ds$$
$$= xe^{txJ_0}I_{2n} - J_0Q(t, t)e^{txJ_0} + Q(t, t)J_0e^{txJ_0} -$$
$$\quad - x\int_t^\infty Q(t, s)J_0e^{sxJ_0}ds + (J_0Q(t, t) - Q(t, t)J_0)Q(t, s))e^{sxJ_0}ds$$
$$= (xI_{2n} + (J_0Q(t, t) - Q(t, t)J_0))U(t, x),$$

where the various integrals converge since the integral

$$\int_t^\infty e^{sT}e^{xsJ_0}ds = \int_0^\infty \begin{pmatrix} e^{-s(a-x)}I_m & 0 \\ 0 & e^{-s(a^*+x)}I_m \end{pmatrix}ds \tag{10.14}$$

converges for all x not in the spectral sets of a or $-a^*$.

STEP 6: *The function* (10.9) *is the only solution of* (10.2) *with the asserted asymptotics.*

We first check that (10.9) satisfies (10.10). Because of the spectra condition on a we have

$$\lim_{t \to \infty} e^{-xt J_0} e^{tT} = 0.$$

On the other hand,

$$\int_t^\infty e^{sT} G e^{s J_0 x} ds = \int_t^\infty \begin{pmatrix} 0 & -e^{-s(a+xI_n)} f^* \\ e^{-s(a^*-xI_n)} c^* & 0 \end{pmatrix} ds$$

$$= \begin{pmatrix} 0 & -(a+xI_n)^{-1} e^{-t(a+xI_n)} f^* \\ (a^*-xI_n)^{-1} e^{-t(a^*-xI_n)} c^* & \end{pmatrix}$$

$$\longrightarrow 0_{2n \times 2n}$$

as $t \to \infty$. It follows that (10.10) is in force. To prove uniqueness consider U_1 and U_2 two solutions. Then $U_1(t, p) \star (U_1(0, p))^{-\star}$ and $U_2(t, p) \star (U_2(0, p))^{-\star}$ have the same initial condition at $t = 0$ and thus coincide. Taking now into account the asymptotic condition (10.10) we get

$$\lim_{t \to \infty} e^{-tx J_0} U_k(t, x) U_k(0, x))^{-1} = U_k(0, x)^{-1}, \quad k = 1, 2,$$

and hence $U_1(0, x)^{-1} = U_2(0, x)^{-1}$. Thus U_1 and U_2 have the same initial condition at $t = 0$, and so $U_1(t, x) = U_2(t, x)$. The result for x replaced by a quaternionic variable follows by slice hyperholomorphic extension. $\qquad\square$

Proposition 10.1.4. *Let $\Theta(t, p)$ be the solution of (10.1) with initial condition $\Theta(0, p) = I_n$. Then, for $x, y \in \mathbb{R} \cap \Omega$ we have:*

$$\int_0^a \Theta(u, x)^* \Theta(u, y) du = \frac{\Theta(a, x)^* J_0 \Theta(a, y) - J_0}{x + y}. \tag{10.15}$$

Proof. We have

$$x \int_0^a \Theta(u, x)^* \Theta(u, y) du + \int_0^a \Theta(u, x)^* \Theta(u, y) y \, du =$$

$$= \int_0^a \left(\frac{d}{du} \Theta(u, x)^* J_0 + \Theta(u, x)^* V(u) \right) \Theta(u, y) du +$$

$$+ \int_0^a \Theta(u, x)^* \left(\frac{d}{du} J_0 \Theta(u, y) - V(u) \Theta(u, y) \right) du$$

$$= \int_0^a \frac{d}{du} \left(\Theta(u, x)^* J_0 \Theta(u, y) \right) du$$

$$= \Theta(a, x)^* J_0 \Theta(a, y) - J_0,$$

and hence the result. $\qquad\square$

Multiplying on the left by $(I_n \ I_n)$ and by its transpose on the right, and setting

$$(I_n \ I_n) \, \Theta(t, x) = \left(E_+(t, x) \ E_-(t, x) \right), \tag{10.16}$$

we get

$$\int_0^a (I_n \ I_n) \, \Theta(u, x)^* \Theta(u, y) \begin{pmatrix} I_n \\ I_n \end{pmatrix} du = \frac{E_+(a, x) E_+(a, y)^* - E_-(a, x) E_-(a, y)^*}{x + y}. \tag{10.17}$$

We set $S(a, p) = E_+(a, p)^{-*} E_-(a, p)$. Then the kernel on the right side of (10.17) can be rewritten as

$$E_+(a, x) \frac{I_n - S(a, x) S(a, y)^*}{x + y} E_+(a, y)^*. \tag{10.18}$$

We set

$$K_{S_a}(x, y) = \frac{I_n - S(a, x) S(a, y)^*}{x + y}. \tag{10.19}$$

Theorem 10.1.5. *For every* $f \in \mathbf{L}_2([0, a], dx, \mathbb{H}^n)$ *there exists a* $Tf \in \mathbf{H}_2(\Pi_+) \ominus S_a \star \mathbf{H}_2(\Pi_+)$ *such that*

$$\int_0^a (I_n \ I_n) \, \Theta(u, p) f(u) du = E_+(a, p)(Tf)(p) \tag{10.20}$$

and the map $f \mapsto Tf$ *is unitary.*

Proof. We now study the image of the operator $J_0 \frac{d}{dt} f - Vf$ under T. From (10.17) we see that the function

$$\Theta(u, y) \begin{pmatrix} I_n \\ I_n \end{pmatrix} c, \quad c \in \mathbb{H}^n,$$

is sent isometrically to the function $K_{S_a}(u, y) E_+(a, y)^* c$, and similarly for finite linear combinations of such functions. For a finite linear combination

$$x(u) = \sum_{m=1}^M \Theta(u, y_m) \begin{pmatrix} I_n \\ I_n \end{pmatrix} c_m$$

the image is

$$\sum_{m=1}^{M} y_m K_{S_a}(u, y_m) E_+(a, y_m)^* c_m = x \left(\sum_{m=1}^{M} y_m K_{S_a}(u, y_m) E_+(a, y_m)^* c_m \right) +$$

$$+ \sum_{m=1}^{M} E(a, y_m)^* c_m -$$

$$- S_a(a, x) \left(\sum_{m=1}^{M} y S_a(a, y_m)^* E(a, y_m)^* c_m \right),$$

that is

$$T x(u) = u x(u) + c_x + S_a(u, a) d_x$$

with

$$c_x = \sum_{m=1}^{M} E(a, y_m)^* c_m,$$

$$d_x = - \sum_{m=1}^{M} y S_a(a, y_m)^* E(a, y_m)^* c_m. \qquad \square$$

10.2 The Characteristic Spectral Functions

The function $U(0, p)$ plays a key role in direct and inverse problems associated with (10.2); following the complex case we will call it the *asymptotic equivalence matrix function*.

Theorem 10.2.1. *The asymptotic equivalence matrix function associated with a potential of the form* (10.3) *is rational and J_0-unitary,*

$$U(0, x) J_0 U(0, -x)^* = J_0, \tag{10.21}$$

at all real points where it is defined, with a realization given by

$$U(0, p) = I_{2m} + C \star (p I_{2m} - A)^{-\star} B \tag{10.22}$$

where

$$A = \begin{pmatrix} a^* & 0 \\ 0 & -a \end{pmatrix} \tag{10.23}$$

$$B = \begin{pmatrix} c^* & 0 \\ 0 & (I_m + \Omega Y)^{-1}(b + \Omega c^*) \end{pmatrix} \tag{10.24}$$

$$C = \begin{pmatrix} -c\Omega & c(I_m + \Omega Y) \\ -(b^* + c\Omega) & (b^* + c\Omega)Y \end{pmatrix}, \tag{10.25}$$

that is,

$$U_{11}(0, p) = I_n - c\Omega(pI_m - a^*)^{-*}c^* \tag{10.26}$$
$$U_{12}(0, p) = c(I_m + \Omega Y)(a + pI_m)^{-*}(I_m + \Omega Y)^{-1}(b + \Omega c^*) \tag{10.27}$$
$$U_{21}(0, p) = -(b^* + c\Omega)(pI_m - a^*)^{-*}c^* \tag{10.28}$$
$$U_{22}(0, p) = I_n + (b^* + c\Omega)Y(a + pI_m)^{-*}(I_m + \Omega Y)^{-1}(b + \Omega c^*). \tag{10.29}$$

Finally the associated Hermitian matrix to the realization (10.23)–(10.25) is

$$H = \begin{pmatrix} -\Omega & I_m + \Omega Y \\ I_m + Y\Omega & -(Y + Y\Omega Y) \end{pmatrix}. \tag{10.30}$$

Proof. We have

$$U(0, p) = I_{2n} + F(I_{2m} - Z)^{-1} \int_0^\infty e^{sT} G e^{sx J_0} ds$$

$$= I_{2n} + \begin{pmatrix} c & 0 \\ 0 & f \end{pmatrix} \begin{pmatrix} I_m & -\Omega(I_m + Y\Omega)^{-1} \\ -Y & I_m \end{pmatrix}^{-1} \times$$

$$\times \int_0^\infty \begin{pmatrix} e^{-sa} & 0 \\ 0 & e^{-sa^*} \end{pmatrix} \begin{pmatrix} 0 & -f^* \\ -c^* & 0 \end{pmatrix} \begin{pmatrix} e^{sx} I_n & 0 \\ 0 & e^{-sx} I_n \end{pmatrix} ds.$$

We have

$$\int_0^\infty \begin{pmatrix} e^{-sa} & 0 \\ 0 & e^{-sa^*} \end{pmatrix} \begin{pmatrix} 0 & -f^* \\ -c^* & 0 \end{pmatrix} \begin{pmatrix} e^{sx} I_n & 0 \\ 0 & e^{-sx} I_n \end{pmatrix} ds$$

$$= \begin{pmatrix} 0 & (a + xI_m)^{-1} f^* \\ (a^* - xI_m)^{-1} c^* & 0 \end{pmatrix}$$

Finally we check that (4.5) and (4.6) are satisfied with H given by (10.30). $\qquad\square$

The following result is the counterpart of [37, Theorem 3.1, p. 14].

Theorem 10.2.2. *There exists a unique $\mathbb{H}^{2n \times n}$-valued solution $X(t, p)$ to the canonical system (10.2) subject to the boundary conditions*

$$\left(I_n \ -I_n\right) X(0, p) \equiv 0, \tag{10.31}$$
$$\lim_{\substack{q=x\in\mathbb{R} \\ t\to\infty}} \left(0 \ e^{tx} I_n\right) U(t, x) = I_n. \tag{10.32}$$

Then the limit

$$\lim_{\substack{q=x\in\mathbb{R} \\ t\to\infty}} \left(e^{-tx} I_n \ 0\right) U(t, x) \tag{10.33}$$

exists.

Proof. The proof is easily adapted from the of [37, Theorem 3.1, p. 14]. We look for a solution of the form $X(t, p) = U(t, p) \star \begin{pmatrix} S(p) \\ L(p) \end{pmatrix}$, where S and L are slice hyperholomorphic. Condition (10.31) gives

$$(U_{11}(0, x) - U_{21}(0, x))S(x) = (U_{22}(0, x) - U_{12}(0, x))L(x).$$

Rewriting (10.32) as

$$\lim_{\substack{q=x\in\mathbb{R} \\ t\to\infty}} \left(0 \; I_n\right) e^{-txJ_0} U(t, x) = 0,$$

and taking into account (10.10) we then obtain $B(x) = I_n$, and hence $L(p) = I_n$, and

$$S(p) = (U_{11}(0, p) - U_{21}(0, p))^{-\star} \star (U_{22}(0, p) - U_{12}(0, p)). \qquad (10.34)$$

□

The function (10.34) is called the *scattering function* associated with the system (10.2).

Theorem 10.2.3. *The scattering function is rational, unitary in the sense that*

$$S(x)S(-x)^* = I_n \qquad (10.35)$$

at all real points where it is defined, and admits a Wiener-Hopf factorization $S(p) = S_+(p) \star S_-(p)$ *where*

$$S_-(p) = I_n - b^* \star (pI_m - a^*)^{-\star}c^*, \qquad (10.36)$$
$$S_+(p) = I_n - (b^*Y - c)(I_m + \Omega Y)^{-1} \star (pI_m + a^\times)^{-\star}(b + \Omega c^*), \qquad (10.37)$$

with inverses

$$S_-(p)^{-\star} = I_n + b^* \star (pI_m - a^{\times*})^{-\star}c^*, \qquad (10.38)$$
$$S_+(p)^{-\star} = I_n + (b^*Y - c) \star (pI_m + a)^{-\star}(I_m + Y\Omega)^{-1}(b + \Omega c^*). \qquad (10.39)$$

Proof. Equation (10.35) is obtained after multiplying both sides of (10.21) by $\left(I_n \; -I_n\right)$ on the left and by its transpose on the right.

The formulas for S_- and $S_+^{-\star}$ follow directly from (10.26)–(10.29). To obtain S_+ we use formula (4.2) for the realization of the inverse; taking into account the Lyapunov equations (10.4) and (10.5) and the definition of $a^\times$ (that is, $a^\times = a - bc$) we first compute

$$-a - (I_m + \Omega Y)^{-1}(b + \Omega c^*)(b^* Y - c)$$
$$= -a - (I_m + \Omega Y)^{-1}(bb^* Y + \Omega c^* b^* Y - bc - \Omega cc^*)$$
$$= -a - (I_m + \Omega Y)^{-1}\{(\Omega a^{\times *} + a^\times \Omega)Y +$$
$$+\Omega(a^* - a^{\times *})Y + a^\times - a - \Omega(Ya + a^* Y)\}$$
$$= -a - (I_m + \Omega Y)^{-1}\{a^\times(I_m + \Omega Y) - (I_m + \Omega Y)a\}$$
$$= -(I_m + \Omega Y)^{-1}a^\times(I_m + \Omega Y). \qquad \qquad \square$$

We note that S also admits a factorization of the type (4.7).

The inverse scattering problem (as opposed to the lossless inverse scattering problem defined earlier) consists in finding the potential associated with a function satisfying the hypothesis of the previous theorem.

We now define two other characteristic spectral functions, namely, the spectral function and the Weyl function. As in the classical case, it will be interesting to solve direct and inverse problems associated with these two functions.

Definition 10.2.4. *The function $S_-^{-*}S_-^{-1}$ is called the spectral function.*

Definition 10.2.5. *Let $\Theta(t, p)$ be the matrizant, defined as the solution to (10.2) with initial condition $\Theta(0, p) = I_{2n}$. The uniquely defined $\mathbb{H}^{n \times n}$-valued function N such that*

$$\int_0^\infty (N(x)^* \; I_n) \, \Theta(u, x)^* \Theta(u, x) \begin{pmatrix} N(x) \\ I_n \end{pmatrix} du < \infty \qquad (10.40)$$

is called the Weyl function.

10.3 Canonical Differential Systems Associated with an Operator

We follow closely the papers [37, 81], and [99–101], and associate to an operator A satisfying (8.1) a canonical differential expression of the type (10.1).

Let A be a right quaternionic operator in the right quaternionic Hilbert space $\mathcal{H}$, with finite dimensional part, and write as in (8.1) above:

$$A + A^* = -C^* J C.$$

Lemma 10.3.1. *Consider the differential equation*

$$D'(t) = -JD(t)A, \quad t > 0 \qquad (10.41)$$
$$D(0) = C, \qquad (10.42)$$

where the unknown $D(t)$ is $\mathbf{L}(\mathcal{H}, \mathbb{H}^n)$-valued and let $\Sigma(t)$ be the $\mathbf{L}(\mathcal{H}, \mathcal{H})$-valued function defined by

$$\Sigma(t) = I_{\mathcal{H}} + \int_0^t D(u)^* D(u) du, \quad t > 0. \tag{10.43}$$

Then

$$\Sigma(t)A + A^*\Sigma(t) = -D(t)^* J D(t), \quad t > 0. \tag{10.44}$$

Proof. Differentiating both sides of (10.44) we obtain

$$D(t)^* D(t)A + A^* D(t)^* D(t) = -D'(t)^* J D(t) - D(t)^* J D'(t),$$

which holds since $D'(t) = -J D(t)A$ and $J^2 = I_n$. Equation (10.44) reduces to (8.1) for $t = 0$, and hence (10.44) holds for all $t \geq 0$. $\qquad\square$

Define

$$S(t, p) = I_n - p D(t)\Sigma(t)^{-1} \star (I_n - pA^*)^{-\star} D(t)^* J. \tag{10.45}$$

Proposition 10.3.2. *The function S in (10.45) is J-contractive for every $t \geq 0$.*

Proof. We first recall that the inverse mapping theorem holds in the quaternionic setting; as in the complex case this follows from the open mapping theorem, still valid in the quaternionic setting; see [23, p. 73]. Hence $\Sigma(t)$ is boundedly invertible since $\Sigma(t) \geq I_{\mathcal{H}}$. We also note that any bounded positive quaternionic operator has a unique positive squareroot; this follows from the spectral theorem for quaternionic self-adjoint operators; see [14]. Rewrite now (10.44) as

$$A(t) + A(t)^* = -C(t)^* J C(t)$$

with

$$A(t) = (\Sigma(t))^{1/2} A(\Sigma(t))^{-1/2} \quad \text{and} \quad C(t) = D(t)(\Sigma(t))^{-1/2}.$$

Then, (10.45) becomes

$$S(t, p) = I_n - p C(t)\Sigma(t) \star (I_n - pA(t)^*)^{-\star} C(t)^* J, \tag{10.46}$$

and the result follows then from Proposition 8.1.1. $\qquad\square$

The following is an adaptation of [81, Proposition 2.2].

Proposition 10.3.3. *With the notation of Lemma 10.3.1, the function $W(t, p) = S(t, p) \star e^{\frac{Jt}{p}}$ is a solution of the canonical differential system*

$$\frac{\mathrm{d}}{\mathrm{dt}} W(t, p) = \left(J H(t) + \frac{J}{p} \right) \star W(t, p) \tag{10.47}$$

where

$$H(t) = D(t)\Sigma^{-1}(t)D(t)^*J - JD(t)\Sigma^{-1}(t)D(t)^*. \tag{10.48}$$

Proof. We follow the proof of [81, Proposition 2.2].

STEP 1: *We find a differential equation satisfied by the function* $D(t)\Sigma(t)^{-1}$.
Removing the dependence on t to lighten the notation we have:

$$\begin{aligned}
(D\Sigma^{-1})' &= -JDA\Sigma^{-1} - D\Sigma^{-1}\Sigma'\Sigma^{-1} \\
&= -JD(-\Sigma^{-1}D^*JD\Sigma^{-1} - \Sigma^{-1}A^*) - D\Sigma^{-1}D^*D\Sigma^{-1} \\
&= J(D\Sigma^{-1})A^* + J(D\Sigma^{-1}D^*J - JD\Sigma^{-1}D^*)(D\Sigma^{-1}).
\end{aligned}$$

STEP 2: *We assume* $p = x \in \Omega \cap (0, \infty)$ *and show the derivative of* $S(t, x)$ *with respect to* t *satisfies*

$$\frac{\mathrm{d}}{\mathrm{d}t}S(t, x) = \left(JH(t) + \frac{J}{x}\right)S(t, x) - \frac{S(t, x)J}{x}. \tag{10.49}$$

The $\star$-product reduces to the pointwise product and we have (still removing the dependence on t in most of the instances)

$$\begin{aligned}
\frac{\mathrm{d}}{\mathrm{d}t}S(t, x) = &- xJ(D\Sigma^{-1})A^*(I - xA^*)^{-1}D^*J- \\
&- xJ(D\Sigma^{-1}D^*J - JD\Sigma^{-1}D^*)(D\Sigma^{-1})(I - xA^*)^{-1}D^*J+ \\
&+ xD\Sigma^{-1}(I - xA^*)^{-1}A^*D^*.
\end{aligned} \tag{10.50}$$

Writing $xA^*(I - xA^*)^{-1} = -I + (I - xA^*)^{-1}$ we can rewrite the first and third terms in the above sum as:

$$\begin{aligned}
-xJ(D\Sigma^{-1})A^*(I - xA^*)^{-1}D^*J &= JD\Sigma^{-1}D^*J - JD\Sigma^{-1}(I - xA^*)^{-1}D^*J \\
xD\Sigma^{-1}(I - xA^*)^{-1}A^*D^* &= -D\Sigma^{-1}D^* + D\Sigma^{-1}(I - xA^*)^{-1}D^*.
\end{aligned}$$

We further remark that

$$\begin{aligned}
JD\Sigma^{-1}D^*J - D\Sigma^{-1}D^* &= JH, \\
-JD\Sigma^{-1}(I - xA^*)^{-1}D^*J &= \frac{J(S(x, p) - I_n)}{x}, \\
-D\Sigma^{-1}(I - xA^*)^{-1}D^* &= \frac{(S(x, p) - I_n)J}{x}.
\end{aligned}$$

Thus (10.50) can be rewritten as:

$$\frac{d}{dt}S(t,x) = JH(S - I_n) + JH +$$

$$+ \frac{J(S - I)}{x} - \frac{(S - I)J}{x} \tag{10.51}$$

$$= (JH + \frac{J}{x})S - \frac{SJ}{x}.$$

It is then clear that:

STEP 3: *The function* $W(t,x) = S(t,x)e^{\frac{Jt}{x}}$ *satisfies*

$$\frac{d}{dt}W(t,x) = \left(JH(t) + \frac{J}{x}\right)W(t,x).$$

□

Remark 10.3.4. The operator A does not determine uniquely H. Indeed, take $a > 0$ and $c = \sqrt{2a}$ so that (8.1) is met with $J = -1$. Then (10.48) leads to $H = 0$ since J is a real scalar, but $S(x) = \dfrac{1 + xa}{1 - xa}$.

Remark 10.3.5. The matrix function H satisfies $H(t) + H(t)^* = 0$. When

$$J = \begin{pmatrix} 0 & I_n \\ I_n & 0 \end{pmatrix} \tag{10.52}$$

it is of interest to find when H is of the form

$$H(t) = \begin{pmatrix} 0 & v(t) \\ -v(t)^* & 0 \end{pmatrix}. \tag{10.53}$$

The $\mathbb{H}^{n \times n}$-valued function v is then called the potential.

We conclude this section with an example.

Example 10.3.6. With A, J, and C as in Example 8.1.2, the differential equation (10.41) has solution

$$D(t) = \begin{pmatrix} 0 & 1 + t \\ 1 & 0 \end{pmatrix},$$

and the function (10.43) equals

$$\Sigma(t) = \begin{pmatrix} 1 & 0 \\ 0 & 1 \end{pmatrix} + \int_0^t D(u)^* D(u)\,du = \begin{pmatrix} 1 & 0 \\ 0 & 1 \end{pmatrix} + \int_0^t \begin{pmatrix} 1 & 0 \\ 0 & (1 + u)^2 \end{pmatrix} du$$

$$= \begin{pmatrix} 1 + t & 0 \\ 0 & \frac{(1+t)^3 + 2}{3} \end{pmatrix}.$$

Hence, for real x

$$S(t, x) = \begin{pmatrix} \dfrac{(1+t)^3 + 2 + 3x^2(1+t)^2}{(1+t)^3 + 2} & \dfrac{3x(1+t)}{(1+t)^3 + 2} \\ x & 1 \end{pmatrix},$$

with associated potential

$$v(t) = \frac{3(1+t)}{(1+t)^3 + 2} - 1.$$

References

1. Kh. Abu-Ghanem, D. Alpay, F. Colombo, D.P. Kimsey, I. Sabadini, Boundary interpolation for slice hyperholomorphic Schur functions. Integral Equ. Oper. Theory **82**, 223–248 (2015)
2. V.M. Adamjan, On the theory of canonical differential operators in Hilbert space. Dokl. Akad. Nauk SSSR **178**, 9–12 (1968)
3. D. Alpay, Some remarks on reproducing kernel Kreĭn spaces. Rocky Mt. J. Math. **21**, 1189–1205 (1991)
4. D. Alpay, Algorithme de Schur, espaces à noyau reproduisant et théorie des systèmes, in *Panoramas et Synthèses [Panoramas and Syntheses]*, vol. 6 (Société Mathématique de France, Paris, 1998)
5. D. Alpay, *An Advanced Complex Analysis Problem Book. Topological Vector Spaces, Functional Analysis, and Hilbert Spaces of Analytic Functions* (Birkhäuser/Springer Basel AG, Basel, 2015)
6. D. Alpay, J. Ball, I. Gohberg, L. Rodman, *Realization and Factorization of Rational Matrix Functions with Symmetries*. Operator Theory: Advances and Applications, vol. 47 (Birkhäuser Verlag, Basel, 1990), pp. 1–60
7. D. Alpay, J. Ball, I. Gohberg, L. Rodman, J-unitary preserving automorphisms of rational matrix functions: state space theory, interpolation and factorization. Linear Algebra Appl. **197–198**, 531–566 (1994)
8. D. Alpay, V. Bolotnikov, F. Colombo, I. Sabadini, Interpolation problems for certain classes of slice hyperholomorphic functions. Integral Equ. Oper. Theory **86**(2), 165–183 (2016)
9. D. Alpay, V. Bolotnikov, P. Loubaton, On tangential H_2 interpolation with second order norm constraints. Integral Equ. Oper. Theory **24**, 156–178 (1996)
10. D. Alpay, M. Bożejko, F. Colombo, D.P. Kimsey, I. Sabadini, Boolean convolution in the quaternionic setting. Linear Algebra Appl. **506**, 382–412 (2016)
11. D. Alpay, P. Bruinsma, A. Dijksma, H.S.V. de Snoo, *Interpolation Problems, Extensions of Symmetric Operators and Reproducing Kernel Spaces I*. Operator Theory: Advances and Applications, vol. 50 (Birkhäuser Verlag, Basel, 1991), pp. 35–82
12. D. Alpay, F. Colombo, J. Gantner, D.P. Kimsey, Functions of the infinitesimal generator of a strongly continuous quaternionic group. Anal. Appl. (Singap.) **15**(2), 279–311 (2017)
13. D. Alpay, F. Colombo, J. Gantner, I. Sabadini, A new resolvent equation for the S-functional calculus. J. Geom. Anal. **25**(3), 1939–1968 (2015)
14. D. Alpay, F. Colombo, D. Kimsey, The spectral theorem for quaternionic unbounded normal operators based on the S-spectrum. J. Math. Phys. **57**(2), 023503, 27 (2016)

© The Author(s), under exclusive license to Springer Nature Switzerland AG 2020
D. Alpay et al., *Quaternionic de Branges Spaces and Characteristic Operator Function*, SpringerBriefs in Mathematics,
https://doi.org/10.1007/978-3-030-38312-1

15. D. Alpay, F. Colombo, D. Kimsey, I. Sabadini, Wiener algebra for the quaternions. Mediterr. J. Math. **13**(5), 2463–2482 (2016)
16. D. Alpay, F. Colombo, D.P. Kimsey, I. Sabadini, The spectral theorem for unitary operators based on the S-spectrum. Milan J. Math. **84**(1), 41–61 (2016)
17. D. Alpay, F. Colombo, I. Lewkowicz, I. Sabadini, Realizations of slice hyperholomorphic generalized contractive and positive functions. Milan J. Math. **83**, 91–144 (2015)
18. D. Alpay, F. Colombo, T. Qian, I. Sabadini, The H^∞ functional calculus based on the S-spectrum for quaternionic operators and for n-tuples of noncommuting operators. J. Funct. Anal. **271**(6), 1544–1584 (2016)
19. D. Alpay, F. Colombo, I. Sabadini, Generalized quaternionic Schur functions in the ball and half-space and Krein-Langer factorization, in *Hypercomplex Analysis: New Perspectives and Applications*. Trends in Mathematics (Birkhäuser/Springer Basel AG, Basel, 2014), pp. 19–41
20. D. Alpay, F. Colombo, I. Sabadini, Krein-Langer factorization and related topics in the slice hyperholomorphic setting. J. Geom. Anal. **24**(2), 843–872 (2014)
21. D. Alpay, F. Colombo, I. Sabadini, Inner product spaces and Krein spaces in the quaternionic setting, in *Recent Advances in Inverse Scattering, Schur Analysis and Stochastic Processes*, ed. by D. Alpay, B. Kirstein. Operator Theory: Advances and Applications, vol. 244 (Birkhäuser, 2015), pp. 33–65
22. D. Alpay, F. Colombo, I. Sabadini, Perturbation of the generator of a quaternionic evolution operator. Anal. Appl. (Singap.) **13**(4), 347–370 (2015)
23. D. Alpay, F. Colombo, I. Sabadini, *Slice Hyperholomorphic Schur Analysis*. Operator Theory: Advances and Applications, vol. 256 (Birkhäuser/Springer, Basel, 2016)
24. D. Alpay, A. Dijksma, J. Rovnyak, H. de Snoo, Schur Functions, Operator Colligations, and Reproducing Kernel Pontryagin Spaces. Operator theory: Advances and Applications, vol. 96 (Birkhäuser Verlag, Basel, 1997)
25. D. Alpay, H. Dym, Hilbert spaces of analytic functions, inverse scattering and operator models, I. Integral Equ. Oper. Theory **7**, 589–641 (1984)
26. D. Alpay, H. Dym, Hilbert spaces of analytic functions, inverse scattering and operator models, II. Integral Equ. Oper. Theory **8**, 145–180 (1985)
27. D. Alpay, H. Dym, On applications of reproducing kernel spaces to the Schur algorithm and rational J-unitary factorization, in *I. Schur Methods in Operator Theory and Signal Processing*. Operator Theory: Advances and Applications, ed. by I. Gohberg, vol. 18 (Birkhäuser Verlag, Basel, 1986), pp. 89–159
28. D. Alpay, H. Dym, Structured invariant spaces of vector valued functions, Hermitian forms, and a generalization of the Iohvidov laws. Linear Algebra Appl. **137**(138), 137–181 (1990)
29. D. Alpay, H. Dym, Structured invariant spaces of vector valued functions, sesquilinear forms, and a generalization of the Iohvidov laws. Linear Algebra Appl. **137**(138), 413–451 (1990)
30. D. Alpay, H. Dym, On reproducing kernel spaces, the Schur algorithm, and interpolation in a general class of domains, in *Operator Theory and Complex Analysis (Sapporo, 1991)*. Operator Theory: Advances and Applications, vol. 59 (Birkhäuser, Basel, 1992), pp. 30–77
31. D. Alpay, H. Dym, On a new class of reproducing kernel spaces and a new generalization of the Iohvidov laws. Linear Algebra Appl. **178**, 109–183 (1993)
32. D. Alpay, H. Dym, On a new class of realization formulas and their applications. Linear Algebra Appl. **241**(243), 3–84 (1996)
33. D. Alpay, I. Gohberg, *On Orthogonal Matrix Polynomials*. Operator Theory: Advances and Applications, vol. 24 (Birkhäuser Verlag, Basel, 1988), pp. 25–46
34. D. Alpay, I. Gohberg, Unitary rational matrix functions, in *Topics in Interpolation Theory of Rational Matrix-valued Functions*, ed. by I. Gohberg. Operator Theory: Advances and Applications, vol. 33 (Birkhäuser Verlag, Basel, 1988), pp. 175–222
35. D. Alpay, I. Gohberg, Inverse spectral problem for differential operators with rational scattering matrix functions. J. Differ. Equ. **118**, 1–19 (1995)
36. D. Alpay, I. Gohberg, Pairs of selfadjoint operators and their invariants. Algebra i Anal. **16**(1), 70–120 (2004)

37. D. Alpay, I. Gohberg, M.A. Kaashoek, A.L. Sakhnovich, Direct and inverse scattering problem for canonical systems with a strictly pseudo-exponential potential. Math. Nachr. **215**, 5–31 (2000)
38. D. Alpay, I. Sabadini, Beurling-Lax type theorems in the complex and quaternionic setting. Linear Algebra Appl. **530**, 15–46 (2017)
39. D. Alpay, B. Schneider, M. Shapiro, D. Volok, Fonctions rationnelles et théorie de la réalisation: le cas hyper-analytique. Comptes Rendus Mathématiques **336**, 975–980 (2003)
40. D. Alpay, M. Shapiro, D. Volok, Rational hyperholomorphic functions in R^4. J. Funct. Anal. **221**(1), 122–149 (2005)
41. D. Arov, H. Dym, J-inner matrix functions, interpolation and inverse problems for canonical systems. V. The inverse input scattering problem for Wiener class and rational $p \times q$ input scattering matrices. Integral Equ. Oper. Theory **43**(1), 68–129 (2002)
42. D.Z. Arov, H. Dym, *Bitangential Direct and Inverse Problems for Systems of Integral and Differential Equations*. Encyclopedia of Mathematics and its Applications, vol. 145 (Cambridge University Press, Cambridge, 2012)
43. H. Bart, I. Gohberg, M.A. Kaashoek, *Minimal Factorization of Matrix and Operator Functions*. Operator Theory: Advances and Applications, vol. 1 (Birkhäuser Verlag, Basel, 1979)
44. T. Boros, A. Sayed, T. Kailath, Structured matrices and unconstrained rational interpolation problems. Linear Algebra Appl. **203**(204), 155–188 (1994)
45. L. de Branges, Some Hilbert spaces of analytic functions I. Trans. Am. Math. Soc. **106**, 445–468 (1963)
46. L. de Branges, The expansion theorem for Hilbert spaces of entire functions, in *Entire Functions and Related Parts of Analysis (Proceedings of the Symposium in Pure Mathematics, La Jolla, CA, 1966)* (American Mathematical Society, Providence, R.I., 1968), pp 79–148
47. L. de Branges, *Hilbert Spaces of Entire Functions* (Prentice-Hall Inc., Englewood Cliffs, N.J., 1968)
48. L. de Branges, *Espaces Hilbertiens de fonctions entières* (Masson, Paris, 1972)
49. L. de Branges, J. Rovnyak, Canonical models in quantum scattering theory, in *Perturbation theory and its Applications in Quantum Mechanics*, ed. by C. Wilcox (Wiley, New York, 1966), pp. 295–392
50. L. de Branges, J. Rovnyak, *Square Summable Power Series* (Holt, Rinehart and Winston, New York, 1966)
51. M.S. Brodskiĭ, *Triangular and Jordan Representations of Linear Operators* (American Mathematical Society, Providence, R.I., 1971). Translated from the Russian by J.M. Danskin, Translations of Mathematical Monographs, vol. 32
52. F. Colombo, J. Gantner, Fractional powers of quaternionic operators and Kato's formula using slice hyperholomorphicity. Trans. Am. Math. Soc. **370**(2), 1045–1100 (2018)
53. F. Colombo, J. Gantner, D.P. Kimsey, *Spectral Theory on the S-Spectrum for Quaternionic Operators*. Operator Theory: Advances and Applications, vol. 270 (Birkhäuser/Springer, Cham, 2018)
54. F. Colombo, I. Sabadini, On some properties of the quaternionic functional calculus. J. Geom. Anal. **19**(3), 601–627 (2009)
55. F. Colombo, I. Sabadini, On the formulations of the quaternionic functional calculus. J. Geom. Phys. **60**(10), 1490–1508 (2010)
56. F. Colombo, I. Sabadini, The quaternionic evolution operator. Adv. Math. **227**(5), 1772–1805 (2011)
57. F. Colombo, I. Sabadini, The $\mathcal{F}$-spectrum and the $\mathcal{SC}$-functional calculus. Proc. R. Soc. Edinb. Sect. A **142**, 479–500 (2012)
58. F. Colombo, I. Sabadini, D.C. Struppa, *Entire Slice Regular Functions*. SpringerBriefs in Mathematics (Springer, Cham, 2016)
59. F. Colombo, I. Sabadini, D.C. Struppa, A new functional calculus for noncommuting operators. J. Funct. Anal. **254**(8), 2255–2274 (2008)
60. F. Colombo, I. Sabadini, D.C. Struppa, *Noncommutative Functional Calculus: Theory and Applications of Slice Hyperholomorphic Functions*. Progress in Mathematics, vol. 289 (Birkhäuser/Springer Basel AG, Basel, 2011)

61. T. Constantinescu, A. Sayed, T. Kailath, Displacement structure and completion problems. SIAM J. Matrix Anal. Appl. **16**(1), 58–78 (1995)
62. P. Dewilde, H. Dym, Lossless chain scattering matrices and optimum linear prediction: the vector case. Int. J. Circuit Theory Appl. **9**, 135–175 (1981)
63. P. Dewilde, H. Dym, Lossless inverse scattering, digital filters, and estimation theory. IEEE Trans. Inf. Theory **30**(4), 644–662 (1984)
64. W.F. Donoghue, *Monotone Matrix Functions and Analytic Continuation*. Die Grundlehren der mathematischen Wissennschaften, vol. 207 (Springer, 1974)
65. N. Dunford, J.T. Schwartz, *Linear operators. Part III*. Wiley Classics Library (Wiley, Inc., New York, 1988). Spectral operators, With the assistance of William G. Bade and Robert G. Bartle, Reprint of the 1971 original, A Wiley-Interscience Publication
66. H. Dym, *J-Contractive Matrix Functions, Reproducing Kernel Hilbert Spaces and Interpolation*. Published for the Conference Board of the Mathematical Sciences, Washington, DC (1989)
67. H. Dym, A Hermite theorem for matrix polynomials, in *Topics in Matrix and Operator Theory (Rotterdam, 1989)*. Operator Theory: Advances and Applications, vol. 50 (Birkhäuser, Basel, 1991), pp. 191–214
68. H. Dym, *Shift, Realizations and Interpolation, Redux*. Operator Theory: Advances and Applications, vol. 73 (Birkhäuser Verlag, Basel, 1994), pp. 182–243
69. H. Dym, A. Iacob. Positive definite extensions, canonical equations and inverse problems, in *Proceedings of the Workshop on Applications of Linear Operator Theory to Systems and Networks Held at Rehovot, June 13–16, 1983*, ed. by H. Dym, I. Gohberg. Operator Theory: Advances and Applications, vol. 12 (Birkhäuser Verlag, Basel, 1984), pp. 141–240
70. H. Dym, H.P. McKean, *Gaussian Processes, Function Theory and the Inverse Spectral Problem* (Academic Press, 1976)
71. H. Dym, D. Volok, Zero distribution of matrix polynomials. Linear Algebra Appl. **425**(2–3), 714–738 (2007)
72. S.G. Gal, I. Sabadini, *Quaternionic Approximation: With Application to Slice Regular Functions*. Frontiers in Mathematics (Birkhäuser/Springer, Cham, 2019)
73. J. Gantner, Operator theory on one-sided quaternionic linear spaces: intrinsic S-functional calculus and spectral operators. To appear in Mem. Am. Math. Soc
74. J. Gantner, A direct approach to the S-functional calculus for closed operators. J. Oper. Theory **77**(2), 287–331 (2017)
75. Y. Genin, P. Van Dooren, T. Kailath, J.M. Delosme, M. Morf, On Σ-lossless transfer functions and related questions. Linear Algebra Appl. **50**, 251–275 (1983)
76. G. Gentili, C. Stoppato, D.C. Struppa, *Regular Functions of a Quaternionic Variable*. Springer Monographs in Mathematics (Springer, Heidelberg, 2013)
77. R. Ghiloni, V. Moretti, A. Perotti, Continuous slice functional calculus in quaternionic Hilbert spaces. Rev. Math. Phys. **25**(4), 1350006, 83 (2013)
78. R. Ghiloni, A. Perotti, Slice regular functions on real alternative algebras. Adv. Math. **226**(2), 1662–1691 (2011)
79. I. Gohberg, G. Heinig, Inversion of finite Toeplitz matrices with entries from a noncommutative algebra. Revue roumaine de mathématiques pures et appliquées **19**, 623–665 (1974)
80. I. Gohberg, G. Heinig, Inversion of finite Toeplitz matrices consisting of elements of a noncommutative algebra [mr0353040], in *Convolution Equations and Singular Integral Operators*. Operator Theory: Advances and Applications, vol. 206 (Birkhäuser Verlag, Basel, 2010), pp. 7–46
81. I. Gohberg, M.A. Kaashoek, A.L. Sakhnovich, Canonical systems with rational spectral densities: explicit formulas and applications. Math. Nachr. **194**, 93–125 (1998)
82. I. Gohberg, L. Lerer, *Matrix Generalizations of M.G. Kreĭn Theorems on Orthogonal Polynomials*. Operator Theory: Advances and Applications, vol. 34 (Birkhäuser Verlag, Basel, 1988), pp. 137–202
83. T. Kailath, A theorem of I. Schur and its impact on modern signal processing, in *I. Schur Methods in Operator Theory and Signal Processing*, ed. by I. Gohberg. Operator Theory: Advances and Applications, vol. 18 (Birkhäuser, Basel, 1986), pp. 9–30

84. T. Kailath, A. Sayed, Displacement structure: theory and applications. SIAM Rev. **37**, 297–386 (1995)
85. M. Kaltenbäck, H. Woracek, Pontryagin spaces of entire functions. I. Integral Equ. Oper. Theory **33**(1), 34–97 (1999)
86. M. Kaltenbäck, H. Woracek, Pontryagin spaces of entire functions. II. Integral Equ. Oper. Theory **33**(3), 305–380 (1999)
87. M. Kaltenbäck, H. Woracek, Pontryagin spaces of entire functions. III. Acta Sci. Math. (Szeged) **69**(1–2), 241–310 (2003)
88. M.G. Kreĭn, Distribution of roots of polynomials orthogonal on the unit circle with respect to a sign-alternating weight. Teor. Funkciĭ Funkcional. Anal. i Priložen. Vyp. **2**, 131–137 (1966) (in Russian)
89. M.G. Kreĭn, H. Langer, Über die verallgemeinerten Resolventen und die charakteristische Funktion eines isometrischen Operators im Raume Π_k, in *Hilbert Space Operators and Operator Algebras (Proceedings International Conference, Tihany, 1970)*, pp. 353–399. North-Holland, Amsterdam, 1972. Colloquia Math. Soc. János Bolyai
90. G. Laville, On Cauchy-Kovalewski extension. J. Funct. Anal. **101**(1), 25–37 (1991)
91. X.-J. Li, The Riemann hypothesis for polynomials orthogonal on the unit circle. Math. Nachr. **166**, 229–258 (1994)
92. A. Lindquist, G. Picci, Infinite-dimensional stochastic realizations of continuous-time stationary vector processes, in *Topics in Operator Theory Systems and Networks (Rehovot, 1983)*. Operator Theory: Advances and Applications, vol. 12 (Birkhäuser, Basel, 1984), pp. 335–350
93. M.S. Livšic, On spectral decomposition of linear nonself-adjoint operators. Mat. Sbornik N.S. **34**(76), 145–199 (1954)
94. I.V. Mikhailova, Weyl matrix circles as a tool for uniqueness in the theory of multiplicative representation of J-inner matrix functions [translated from *analysis in infinite-dimensional spaces and operator theory (russian)*, 101–117, "Naukova Dumka", Kiev, 1983], in *Topics in Interpolation Theory (Leipzig, 1994)*. Operator Theory: Advances and Applications, vol. 95 (Birkhäuser, Basel, 1997), pp. 397–417
95. V.P. Potapov, The multiplicative structure of J-contractive matrix–functions. Trudy Moskow. Mat. Obs. **4**, 125–236 (1955). English translation in: American Mathematical Society Translations (2), vol. 15, pp. 131–243 (1960)
96. P. Rao, P. Dewilde, *System Theory for Lossless Wave Scattering*. Operator Theory: Advances and Applications, vol. 19 (Birkhäuser Verlag, Basel, 1986), pp. 333–358
97. L. Rodman, *Topics in Quaternion Linear Algebra*. Princeton Series in Applied Mathematics (Princeton University Press, Princeton, NJ, 2014)
98. G.C. Rota, On models for linear operators. Commun. Pure Appl. Math. **13**, 469–472 (1960)
99. A. Sakhnovich, Exact solutions of nonlinear equations and the method of operator identities. Linear Algebra Appl. **182**, 109–126 (1993)
100. A. Sakhnovich, Canonical systems and transfer matrix-functions. Proc. Am. Math. Soc. **125**(5), 1451–1455 (1997)
101. L.A. Sakhnovich, Factorization problems and operator identities. Russ. Math. Surv. **41**, 1–64 (1986)
102. L.A. Sakhnovich, *Spectral Theory of Canonical Differential Systems. Method of Operator Identities*. Operator Theory: Advances and Applications, vol. 107 (Birkhäuser Verlag, Basel, 1999). Translated from the Russian manuscript by E. Melnichenko
103. L.A. Sakhnovich, *Spectral Theory of Canonical Differential Systems. Method of Operator Identities*. Operator Theory: Advances and Applications, vol. 107 (Birkhäuser Verlag, Basel, 1999). Translated from the Russian manuscript by E. Melnichenko
104. L.A. Sakhnovich, *Integral Equations with Difference Kernels on Finite Intervals*. Operator Theory: Advances and Applications, vol. 84 (Birkhäuser/Springer, Cham, 2015). Second edition, revised and extended
105. A. Sayed, T. Constantinescu, T. Kailath, Time-variant displacement structure and interpolation problems. IEEE Trans. Autom. Control **39**(5), 960–976 (1994)

106. A. Sayed, T. Kailath, Fast algorithms for generalized displacement structures and lossless systems. Linear Algebra Appl. **219**, 49–78 (1995)
107. L. Schwartz, Sous espaces hilbertiens d'espaces vectoriels topologiques et noyaux associés (noyaux reproduisants). J. Anal. Math. **13**, 115–256 (1964)
108. Y. Shelah, Quaternionic Wiener algebras, factorization and applications. Adv. Appl. Clifford Algebras **27**(3), 2805–2840 (2017)
109. Y.L. Shmul'yan, Division in the class of J-expansive operators. Math. Sb. **116**, 516–525 (1967)
110. F. Zhang, Quaternions and matrices of quaternions. Linear Algebra Appl. **251**, 21–57 (1997)

Index

© The Author(s), under exclusive license to Springer Nature Switzerland AG 2020
D. Alpay et al., *Quaternionic de Branges Spaces and Characteristic Operator Function*, SpringerBriefs in Mathematics,
https://doi.org/10.1007/978-3-030-38312-1

Printed in the United States
By Bookmasters